Polar Bears in Northwest Greenland

MONOGRAPHS ON GREENLAND
VOLUME 351

MAN & SOCIETY
VOLUME 41

Polar Bears in Northwest Greenland

An Interview Survey about the Catch and the Climate

Erik W. Born, Anna Heilmann,
Lene Kielsen Holm, Kristin L. Laidre

MUSEUM TUSCULANUM PRESS
UNIVERSITY OF COPENHAGEN
2011

Erik W. Born, Anna Heilmann, Lene Kielsen Holm, Kristin L. Laidre.
Polar Bears in Northwest Greenland: An Interview Survey about the Catch and the Climate

Consultant: Christian Glahder
Translation by Lucinda Ellis
Cover design: Erling Lynder
Set and printed in Denmark by Special-Trykkeriet Viborg a-s
ISBN 978 87 635 3168 9

Monographs on Greenland | Meddelelser om Grønland, vol. 351
ISSN 0025 6676

Man & Society, vol. 41
ISSN 0106 1062
Series editor: Hans Christian Gulløv

www.mtp.dk/MoG

Cover photos:
Front: Tracks of a female polar bear and her yearling cub in Qimmuseriarsuaq/ Melville Bay. Polar bear cubs accompany their mother until they are a little more than two years old. During that period they learn how to hunt and gradually become independent. Photo: Kristin Laidre.
Back: Polar bear footprint. Photo: Øystein Wiig.

This book is published with financial support from
Greenland Institute of Natural Resources

Museum Tusculanum Press
126 Njalsgade
DK-2300 Copenhagen S
Denmark

Acknowledgements

We would like to thank the hunters and residents of the Upernavik and Qaanaaq municipalities for welcoming Anna Heilmann and Tukummeq Qaavigaq with such hospitality, and for the generosity they showed in sharing their knowledge and allowing it to be a part of this study. We wish to thank Tukummeq Qaavigaq for her enthusiastic participation in the study. We also wish to thank KNAPK for their invaluable cooperation – and the local politicians as well as the employees in the municipal offices in Qaanaaq and Upernavik for providing office space and for giving practical assistance in a number of other ways. We would also like to thank the wildlife officers in the Upernavik municipality – Niels Hansen and Otto Eliassen – for their invaluable assistance, Dr. Leif Toudal Pedersen (Danish Meteorological Institute) for the data regarding ice analyses and biologist Signe Sveegaard (Aarhus University and National Environmental Research Institute–Arctic Environment, Roskilde) who helped to organise and analyse some of the data. We also wish to thank Lucinda Ellis for translating the report into English. Ole Heinrich, Elmer Topp-Jørgensen (Department of Fisheries, Hunting and Agriculture, DFHA, Nuuk) and Fernando Ugarte (DFHA, and now Greenland Institute of Natural Resources, Nuuk) assisted us a great deal by providing hunting statistics and other information about the polar bear catch. The study was financed by the Department of Fisheries, Hunting and Agriculture and Pinngortitaleriffik/Greenland Institute of Natural Resources. Thanks to Jens Danielsen, Tukummeq Qavigaq, Mikkel Villum Jensen and Øystein Wiig for allowing us to use their photographs in this report. We wish to thank Dr. Christian Glahder (National Environmental Research Institute, Dept. of Arctic Environment, Aarhus University) who reviewed this paper and offered useful criticism that improved the work. Finally we wish to thank the editor Jordy Findanis for his great help during various phases of the production of this book.

Abstract

In February 2006, 72 experienced polar bear hunters living in the Qaanaaq (25) and Upernavik (47) municipalities (Northwest Greenland) were interviewed. Hunters were asked about polar bear biology, polar bear catch, climate change, and the effect of these changes on the polar bears and the catch. The rationale for this study was the indication from hunting statistics, satellite telemetry and second hand reports that the catch of polar bears in Northwest Greenland had increased since the early 1990s simultaneously with marked changes in weather conditions, sea ice cover and glaciers. The majority of the informants noted an increased occurrence of bears closer to the coast (*i.e.* in areas usually used for hunting). About 31% of the answers specified that the reason for this change was an increase in the number of polar bears, whereas 16% of the answers specified it was due to a decrease in sea ice cover. The hunters in Qaanaaq were more inclined to believe that a decrease in sea ice cover explained the increase in coastal occurrence of polar bears, whereas the informants in Upernavik municipality primarily believed it was an increase in the total number of bears. It was not clear from the results when the change in bear density occurred. However, several informants expressed the opinion that polar bears were scarce during the 1960s and 1970s. During recent years, perhaps beginning in the 1990s, the hunters noted marked environmental changes. Most pronounced, and of greatest importance to hunting, was the decrease in sea ice cover. Ice formed later in the fall and broke up earlier in the spring. Sea ice was also reported to have become thinner. Furthermore, hunters also reported warmer and more unpredictable weather, including frequent storms and rain during winter. Hunters also noted the glaciers retreated and parts of the edge of the Inland Ice Cap recessed. These changes have influenced both travelling and hunting activities in Northwest Greenland, in particular in the municipality of Qaanaaq where previous sled routes along glaciers and sea ice can no longer be used. In contrast, the boating season has been extended with increased open water, resulting in an increase in the fraction of polar bears caught from a boat (versus sled). This change, most pronounced in the southern parts of the Upernavik municipality, was mentioned by some of the informants as an additional reason for the increased catch of bears since the early 1990s (a boat has a larger range and can cover more ground faster than a sled). About 24% of the informants said that polar bears demonstrated physical changes (*e.g.* had become thinner either as a result of increased competition or access to less food due to a decrease in sea ice). Thinner bears were most frequently reported in the Qaanaaq municipality than in the Upernavik municipality (ca. 52% and ca. 10% of the interviewees, respectively). The reason for the regional difference is not clear, but the results may (to an unknown extent) reflect the fact that informants in Qaanaaq generally had more experience hunting polar bears than those in Upernavik. No informants suggested that polar bear foraging had changed. In addition the study presents a variety of observations about the catch of polar bears, polar bear behaviour, and biology. The demography of the catch of polar bears in Northwest Greenland during 1952–2005 is described on the basis of 588 catches (754 polar bears).

Contents

Introduction

The polar bear – *nanoq* – (*Ursus maritimus*) is an animal of great traditional significance (*e.g.* Rasmussen 1919a, b, 1921) in the Qaanaaq and Upernavik municipalities in Northwest Greenland, as well as in the rest of Greenland. Bear hunting provides meat and skins and products such as the hide, claws and skull can be sold and thereby contribute to an otherwise strained hunting economy (*e.g.* Rasmussen 2005). Moreover, the polar bear and the act of bear hunting are traditionally important elements of the mythical universe, the identity of the hunting culture, and the individual hunter (Rasmussen 1919a, b, Born 2008).

Since the early 1990s the polar bear catch in Greenland has increased markedly, with the most notable increase found in the Upernavik municipality (Born & Sonne 2006; Fig. 1). The increased catch has inevitably raised questions as to the underlying reasons since it has coincided with large-scale changes in the natural environment in Greenland (Comiso & Parkinson 2004, ACIA 2005). Data obtained from satellites suggest a decline in the extent of sea ice in the Baffin Bay region (UNEP 2007, Perovich & Richter-Menge 2009), and in particular the eastern regions of Baffin Bay since 2000 (Born & Sonne 2006, Heide-Jørgensen *et al.* 2008, this study). Furthermore, the sea ice in Baffin Bay breaks up earlier each year (Stirling & Parkinson 2006). Various pieces of information have indicated that the polar bear catch in Northwest Greenland has been affected by these environmental changes (*e.g.* Ehrlich 2006).

Population inventories suggest that polar bear populations in Ikersuaq/Kane Basin and Baffin Bay have been over-exploited for a number of years (*e.g.* Aars *et al.* 2006). This conclusion was based on data from population surveys estimating the number of animals in Kane Basin and Baffin Bay (Taylor *et al.* 2003, 2008, Aars *et al.* 2006). In combination with these data, there has been a relatively large combined Canadian-Greenlandic polar bear catch and a marked increase in the reported catch in the municipality of Upernavik (Born & Sonne 2006).

Hence, this resulted in a discrepancy between an increased catch (perhaps indicating an increased occurrence of bears in the hunting areas) and decreasing population sizes in Kane Basin and Baffin Bay. A similar apparent discrepancy was found for polar bears in southwestern Hudson Bay (Tyrrell 2006).

In the case of Northwest Greenland, there was a need to gather information on the polar bear hunters' personal observations and their understanding of the present conditions. This is because bear hunters in the Qaanaaq and Upernavik municipalities have an understanding of the occurrence of animals and changes to the environment given their intimate relationship with their surroundings. Furthermore, a systematic compilation of local observations can serve to supplement scientific biological investigations. Alternatively as said by Usher (2000:187): "It makes good sense to involve people who spend a lot of time on the land in environmental assessment and management, for the obvious reason that they get to see things more often, for longer, and at more different times and places than is normally the case for scientists."

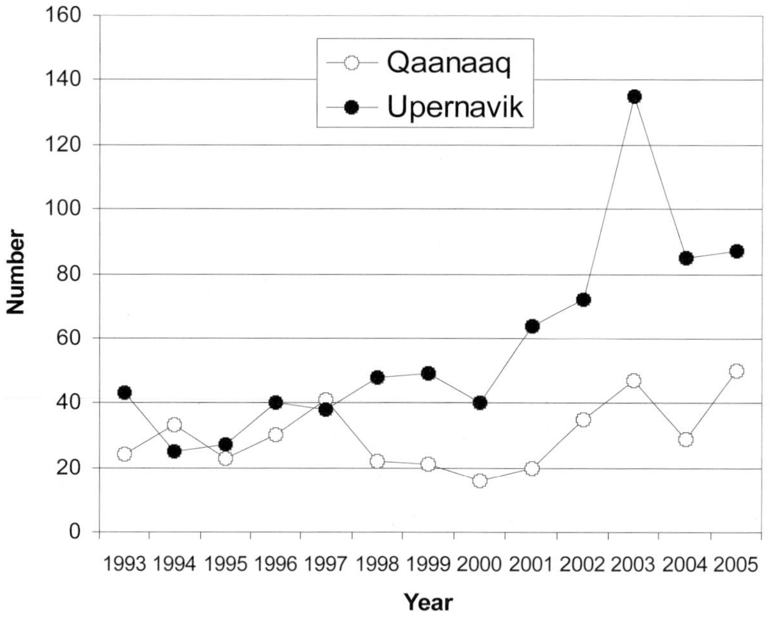

FIG. 1. The polar bear catch reported in the catch recording system *"Piniarneq"* in the Qaanaaq and Upernavik municipalities between 1993 and 2005 (Source: Department of Fisheries, Hunting and Agriculture, Nuuk, 2007). The increase in the catch was statistically significant in the Upernavik municipality ($r^2 = 0.631$, $z = 3.426$, $p = 0.0006$, $n = 13$), but not in the Qaanaaq municipality ($r^2 = 0.159$, $z = 1.334$, $p = 0.182$, $n = 13$).

Interviews with hunters and other locals have been used in a number of studies of living resources in the Arctic (*e.g.* Ferguson & Messier 1997, Usher 2000, Riedlinger & Berkes 2001, Nichols *et al.* 2004 and references therein, Keith *et al.* 2005, Dowsley & Wenzel 2008 and references therein). A recent survey – one which is particularly relevant to this study – gathered information on changes to the physical environment and to the occurrence and catch of polar bears in Nunavut in Canada based on interviews with 48 hunters, elders and community members from the settlements of Mittimatalik/Pond Inlet, Kangiqtugaapik/Clyde River, and Qikiqtarjuaq/Broughton Island on the east coast of Baffin Island, which borders Baffin Bay (Dowsley 2005, Dowsley & Taylor 2006, Dowsley & Wenzel 2008). Local knowledge is often referred to as "traditional ecological knowledge" (TEK) or in case of Nunavut *"Inuit qaujimajatuqangit"* (*e.g.* Dowsley & Wenzel 2008) or *"Inuit qaujimaninangit"* (Keith *et al.* 2005) – or IQ.

In Greenland there have been several studies that have gathered local knowledge regarding the occurrence and catch of a variety of animals, including polar bears (Born 1983, Sandell & Sandell 1991, 1996, Petersen 1993a, b, Glahder 1995, 2001, 2003, Dietz *et al.* 2001, Sandell *et al.* 2001, Aastrup *et al.* 2005). Vibe (1968), Haller (1978), Born (1987), Rosing-Asvid & Born (1990) and Rosing-Asvid (2002) also described aspects of polar bear hunt in Northwest Greenland based on locally gathered information. Rosing-Asvid & Born's interview survey is of particular relevance to the present study because it was conducted in 1989 and 1990 with 70 hunters in the Qaanaaq municipality and 21 in the Upernavik municipality, providing a good basis for comparion to the current study. Rosing-Asvid (2002) used the information in Rosing-Asvid and Born (1990) – supplemented with data from biological samples from the catch (1988–1996) and catch statistics – as the basis for a description of the polar bear catch in the two municipalities.

In February 2006, the Greenland Institute of Natural Resources (Pinngortitaleriffik, Nuuk), in collaboration with the Inuit Circumpolar Council (ICC, Nuuk), conducted an interview survey among polar bear hunters in the municipalities of Qaanaaq and Upernavik (Fig. 2). Seventy-two polar bear hunters were interviewed: 25 in the Qaanaaq municipality and 47 in the Upernavik municipality. The objective was to gather as much information as possible about the hunters' observations of climatic and physical changes in the environment, and how these changes have influenced the polar bear catch. Furthermore, information was sought about the geographic distribution of polar bears as well as the size, age and sex composition of the catch. During the course of this study, a wealth of information was gathered about polar bears in Northwest Greenland and how they are hunted, as well as observations on climate change and a number of other topics. This report summarises the results from the interview survey in 2006 and presents a general overview of trends in the responses while including as many specific observations as possible. The goal was to achieve a reasonable balance between the need to generalise information without excluding the most interesting details.

The report introduces and describes the interview methods and then summarises the responses to the numerous interview questions in three different regions (*i.e.* the Qaanaaq municipality and the northern and southern parts of the Upernavik municipality). A subsequent section presents the geographic distribution and the age and sex composition of the polar bear catch based on data from 588 bear hunts – during which 754 animals were taken between 1952 and 2006. Finally, the results of the study are put into perspective and discussed. In addition to English, the report is available in Greenlandic and Danish (Born *et al.* 2008a, b).

During the time this study was conducted (2006), Greenland was organised into 18 administrative units called 'municipalities'. However, in January 2009 Greenland was re-organised into four major administrative units (Anon. 2008), in which the former Qaanaaq and Upernavik municipalities became part of Qaasuitsup Kommunia ("North Greenland"). However, the interviews reported in this study were conducted prior to the re-organisation and the former municipality units are used to facilitate comparison with earlier studies (*e.g.* Rosing-Asvid & Born 1990).

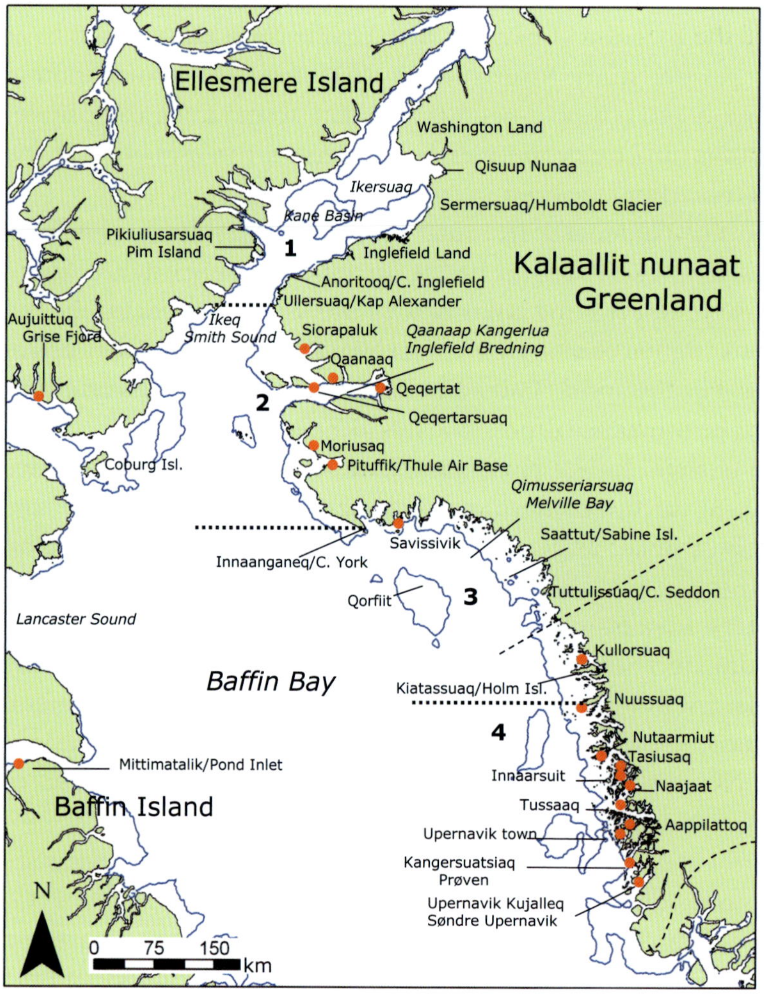

FIG. 2.
The locations of towns and settlements in the Qaanaaq and Upernavik municipalities, together with important names that are mentioned in the text. The boundaries (bold and dotted lines) of the sub-areas (1–4) used in this study for description of the polar bear hunt, municipal boundaries (*i.e.* until 1 January 2009, see Introduction; narrow and dotted) and 200 metre bathymetry curves are shown.

Materials, methods and background

Preparation and selection of informants

Prior to initiating the interviews, the general public and the residents of the Qaanaaq and Upernavik municipalities were informed about the background, character and purpose of the study. The study was mentioned in several Greenlandic language radio programmes, a letter was sent to inform KNAPK (*Kalaallit Nunaanni Aalisartut Piniartullu Kattuffiat, i.e.* the Organization of Fishermen and Hunters in Greenland), and polar bear hunters in the municipalities were informed via a hard copy letter sent with the help of the local branches of KNAPK. The local branches of KNAPK also posted the letter on bulletin boards outside the municipal offices in the settlements. There was therefore a certain amount of anticipation surrounding the interview study before the interviewers arrived in the settlements. In addition, employees from the municipal office in the town Upernavik made phone calls to inform the hunters in the different settlements in the municipality.

The choice of which settlements to visit and the selection of the interviewees was partly based on lists of hunters who had reported polar bear catches to the Department of Fisheries, Hunting and Agriculture (DFHA, the Greenland Home Rule Government, Nuuk) in the preceding 5 years and partly based on where active or formerly active bear hunters lived. These hunters were identified by other local hunters, employees at the municipal offices and KNAPK representatives.

Thus the informants were, for the most part, hunters with specific experience in polar bear hunting. All of the interviewees were men. An attempt was made to interview as many individuals as possible, but the total number was limited by factors such as travel, time constraints, and economic considerations.

A total of 72 hunters were interviewed for the study (25 in the Qaanaaq municipality and 47 in the Upernavik municipality). However, a problem with the recording instruments resulted in one of the interviews from the municipality of Upernavik being incompletely documented (in Nutaarmiut) with another not documented at all (in Nuussuaq); Table 1.

Municipality	Town/ Settlement	Sub-area no.	Number of interviewees	Number of hunters [1)]	Population [2)]
Qaanaaq	Siorapaluk	1	6	14	84
	Qaanaaq	2	9	38	590
	Qeqertarsuaq	2	2	1	2
	Savissivik	3	8	13	77
Upernavik	Kullorsuaq	3	11	55	401
	Nuussuaq	3	7	50	216
	Nutaarmiut	4	3	3	46
	Innaarsuit	4	9	44	149
	Naajaat	4	1	20	65
	Tussaaq	4	1	1	1
	Aappilattoq	4	8	38	196
	Upernavik	4	7	95	1099

Table 1. Distribution according to town/settlement and sub-area (from north to south) of the 72 polar bear hunters interviewed in February 2006 in Qaanaaq (n = 25) and Upernavik (n = 47) municipalities. The number of registered hunters and total population are also shown. Note that hunters from Savissivik, Kullorsuaq and Nuussuaq hunt polar bears in the same region (*i.e.* the Melville Bay). 1: Registered in early 2006; source: Department of Fisheries, Hunting and Agriculture (DFHA, Nuuk). 2: Source: Statistical Year Book – Greenland 2005.

Travel activity

From 8 to 21 February 2006, one of the interviewers (Tukummeq Qaavigaq, TQ) travelled to the Qaanaaq municipality visiting the town of Qaanaaq and three settlements (Siorapaluk, Qeqertarsuaq, Savissivik) where there are active bear hunters (Table 1; Fig. 2). TQ was raised in the municipality of Qaanaaq and therefore spoke and understood the local Greenlandic dialect. Between 2 and 7 February 2006 and again from 15 to 22 February the second interviewer (Anna Heilmann, AH), who comes from Maniitsoq, toured the municipality of Upernavik, visiting the town of Upernavik and five settlements (Kullorsuaq, Nuussuaq, Nutaarmiut, Innaarsuit, Aappilattoq). In addition to the local hunters, hunters who were visiting these settlements from two other settlements (Naajaat, Tussaaq) were also interviewed (Table 1). The timing of the interviews was considered favourable, as the winter darkness and severe cold in February limited hunting activity and there was therefore a relatively high likelihood of finding the informants at home.

The interviewers did not visit the settlements of Qeqertat and Moriusaq (municipality of Qaanaaq) which were home to two and one registered hunters, respectively (source: DFHA). In the Upernavik municipality, the settlements Kangersuatsiaq (Prøven) and Upernavik Kujalleq (Søndre Upernavik) were not visited partly due to travel constraints and partly due to previous information about relatively limited polar bear hunting activity. In these areas there are 45 and 40 registered hunters, respectively (source: DFHA); Fig. 2.

The form and content of the interviews

The interviews were conducted as conversations in Greenlandic (two markedly different dialects in the two municipalities), during which the interviewer asked a series of predetermined questions (Appendix 1). The questions and the questionnaire were prepared by one of the biologists involved in the study (EWB) in collaboration with the interviewers (TQ, AH) and one of the authors (LKH of the Inuit Circumpolar Council, ICC).

The interviews were recorded on cassettes and typically lasted a couple of hours, although sometimes varied greatly in duration. At the start of each interview, all of the bears caught by the hunter were detailed and information was noted down on a special form. Information on the date of each catch, the bear's sex, estimated age, the location of the catch, the names of any others who had taken part in the catch, and the hunting technique were all recorded. The informants were asked to mark the site of each catch on a map. After this, the remaining questions were posed to the informants (Appendix 1), who were also asked to indicate on a map the sites where they had observed changes to the physical environment or other phenomena which could be relevant to the study (*e.g.* observations of dens or hunting routes etc.).

The hunters were asked to categorise each of the bears that they had taken using the following age categories: "old" (*utoqqaq*); "adult" (*inersimasoq*); "juvenile" (*inuusuttoq*) and "cubs" (*piaraq* singular/*piarat* plural) (*i.e.* 1–2-year-old bears). For larger cubs (2–3-year-olds), the informants were asked to specify whether the cubs were with their mother (*i.e.* in family groups) or whether they were alone. Young or juvenile bears that do not accompany their mother were considered to be "independent" or "adult" (Taylor *et al.* 1987, 1988).

During the course of the interviews, a differentiation was made between whether a bear was caught during a hunt using a dogsled, a skiff (usually a 16 foot skiff with a 40 Hp or more powerful outboard motor), or a cutter (pre 1990s 20–25 foot wooden cutters and after the early 1990s fibreglass cutter type AWI 27; www.awi.fo). Bears could also be categorised as animals that had wandered on their own into a town, settlement, or areas that are otherwise regularly visited by hunters (*e.g.* places where seal nets are set).

The hunters were guaranteed anonymity and are therefore not named in this report. It is however unavoidable that, in specific cases where some individuals have an intimate knowledge of the local community, it may be possible to identify an informant based on the details provided by that informant. All of the informants received 600 DKK (ca. $ 100 US, May 2010) in accordance with the accepted rates as compensation for lost earnings given the fact they were asked to stay home to be interviewed even if it was good hunting weather.

Analysis of the data

The two interviewers translated and transcribed all interviews as accurately as possible. The interviews from the Qaanaaq municipality were translated directly into Danish by TQ, while the remaining interviews by AH were transcribed into West Greenlandic and Danish, respectively.

A database was created containing all of the information from each individual catch. This database contained information about (1) the identity of the hunter and settlement, (2) the date, year and location of each catch, (3) the estimated age of the animal, (4) whether the animal was part of a family group (number of young and their sex and age), (5) whether the young were shot, (6) the names of any other hunters who took part in the hunt, (7) the hunting technique (sled, skiff, cutter or whether the animal walked into populated areas on its own accord) and (8) whether the animal was tagged (ear tags, satellite radio, lip tattoos). The data from the questionnaire forms were cross-checked with the oral information given by the hunters. Cases of multiple reporting were discovered by comparing data on the year, month, location, the sex and age of the animal, as well as the names of the other hunters who took part in the hunt. Incidents of double reporting were eliminated in the subsequent presentation of data and information (see the chapter "The catch in figures", p. 180).

A second database contained information (in the form of keywords) about how each of the hunters answered each question. A third database provided an overview of the maps on which the hunters marked phenomena such as breeding areas, denning areas, etc.

Hunting areas

The Qaanaaq municipality stretched from the Petermann Glacier in Hall Basin (about 81° N) to the northernmost border of the Upernavik municipality at 75° 00' N in Qimusseriarsuaq/Melville Bay. Polar bear hunting in the Qaanaaq municipality did not take place any further north than Nuussuaq/Washington Land in Ikersuaq/ Kane Basin at around 80° N (Rosing-Asvid & Born 1990, this study). The Upernavik municipality stretched from 75° 00' N to 72° 01' N at Nuussuaq (Svartenhuk Peninsula) (Anon. 1985, Jakobsen *et al.* 2000); Fig. 2.

In order to describe the hunting activities, the municipalities of Qaanaaq and Upernavik were divided into four sub-areas based on differences in ice coverage and current patterns (Areas 1–4; Fig. 2). The Qaanaaq municipality was sub-divided into a northern section (Area 1) which encompassed the area to the north of Ullersuaq/ Kap Alexander – 78° 10' N (and concerning bear hunting in the past, also comprises the eastern part of Umimmaat Nunaat/Ellesmere Island in Canada) and a central section (Area 2) that consisted of the area between Ullersuaq and Innaanganeq/ Kap York (76° 30' N). Area 3 encompassed Qimusseriarsuaq/Melville Bay, to the south of Innaannganeq/Kap York. Area 3 also included the northern Upernavik area between Innaanganeq/Kap York and Nuussuaq (Kraulshavn, ca. 74° N). Area 4 comprised the Upernavik municiplaty to the south of Nuussuaq (Fig. 2).

This division was partly due to variation in the physical environment in the two areas and partly due to the fact that hunters from Nuussuaq and Kullorsuaq predominantly hunt polar bears north of 74° N. About 88% of 93 catches in Upernavik municipality in the areas to the north of 74° N were made by hunters from Nuussuaq and Kullorsuaq. Out of a total of 182 catches from areas to the south of 74° N, only ca. 2.7% were reported by hunters from Nuussuaq. The Upernavik catches which occurred in the area to the north of 74° N were therefore taken in Area 3 in the presentation of information from the municipality of Qaanaaq. Results from Area 3 in the Qaanaaq municipality and Area 3 in the Upernavik municipality are therefore presented together in certain contexts.

Time periods

In several cases, we present the information for two time periods: Before 1990 and after 1990 (*i.e.* 1952–1990 and 1991–2005) or sometimes for the 5-year period between 2001 and 2005 (data from January and February 2006 were not included). There are several reasons for this: (1) The majority of the information in this study regarding the bear harvest dates from the last 15 years, and especially from 2001–

2005, (2) since the early 1990s and particularly between 2001–2005 there has been an increase in the air temperature in Northwest Greenland (Cappelen 2006) and in the sea temperature in West Greenland (Riebergaard & Buch 2005, Riebergaard 2006, Holland *et al.* 2008), (3) satellite-based sea ice information indicates that there has been a reduction in sea ice cover since 2000 (Born & Sonne 2006) and that in the same period the sea ice has been forming later and breaking up earlier in the north-eastern part of Baffin Bay (*i.e.* the area along the coast of the Upernavik municipality) (Stirling & Parkinson 2006, this study), (4) the past 15 years cover the period of time since the introduction of a new system of reporting catches, called "*Piniarneq*" (introduced in 1993; Kapel & Rosing-Asvid 1996), (5) regulations governing the polar bear hunt in Greenland and in the Melville Bay Nature Reserve, as well as the rules preventing Greenlanders from hunting polar bears on Canadian territory, were all introduced before 1991, (6) many of the statements made by the hunters indicated that significant changes to the natural environment took place during the 1990s; and finally, (7) the previous interview surveys of the polar bear catch in the Qaanaaq and Upernavik municipalities were conducted in 1989 and 1990, respectively (Rosing-Asvid & Born 1990).

For certain aspects of this analysis, the year was divided into three "ecological seasons" (Born *et al.* 2004): "winter" (1 November–31 March) or the period when fast ice typically covers the fjords and darkness prevails for the majority of the time, "spring" (1 April–31 July) when there is daylight and ringed seals (*Phoca hispida*) raise pups in snow lairs and increasingly spend time basking on the ice as "*uuttut*" (singular: *uuttoq*; *i.e.* the term for seals that bask on the ice) until the sea ice breaks up, and "summer" or the "open water season" (1 August–31 October) between fast ice breakup and the formation of new ice in the fjords. These divisions are recognised to be somewhat inflexible considering the changes in breakup of the sea ice and formation of new ice in recent years (this study). Furthermore, it is recognised that the timing of the phenomena detailed above may be different in the southern reaches of the Upernavik municipality and in the north of the Qaanaaq municipality.

Terminology and statistical analyses

The term "bear catch" is used in this report to describe cases where one or more polar bears were killed (this could be an individual bear or a family group, *i.e.* a female bear with cubs). Otherwise, where relevant, we specify the exact number of individual bears that were killed.

The terms "adult" or "independent" are used to describe adult female polar bears as well as all other polar bears that were not a member of a family group (Taylor *et al.* 1987).

χ^2-statistics were used to compare the difference in proportion among areas or periods of time. Pearson's correlations were used to evaluate the relationship between temperature and time. We used the level of significance $P = 0.05$.

In some cases a question was missed during an interview and in others we subsequently discovered that there had been technical problems with the recording equipment. Hence, the number in the tables listing the number of responses may vary.

In several cases we attempted to categorise the answers according to their content and present the categories in tables. Because some answers contained information belonging to more than one category, the number of answers in these tables can be higher than the number of informants asked.

In the verbatim quotes from the hunters, "[........]" means that, for the sake of clarity, a part of the text was omitted which was not relevant to the overall meaning. Curved brackets are used to indicate instances where we have commented on a statement or an expression (*e.g.* author's note).

We used ArcView (Version 9.2, ESRI, Redlands, Ca., USA) to create maps of the geographical distribution of the catch. Statistical analyses were carried out using Statview for Windows 4.5 (Abacus Concepts, 1992–1998) and Excel 2000 (Microsoft).

Background

In order to provide the reader a background for evaluating the results of the survey, we briefly describe the physical factors that affect the occurrence of polar bears and hunting activity in Northwest Greenland. The regulations governing polar bear hunting in Greenland have also (since their introduction in 1974) influenced hunting patterns as they limit the times of the year and areas where hunting can take place and determine which bears may be killed according to age and sex. We therefore also briefly summarise the polar bear hunting regulations in Greenland. A presentation of changes in temperatures and sea ice in the Qaanaaq and Upernavik municipalities is also pertinent to the evaluation of the information from the interviewees on environmental changes in the hunting areas.

Physical factors

The distribution and catch of polar bear in Northwest Greenland depends largely on three main factors (Rosing-Asvid & Born 1990): (1) the distribution of reasonably

The glacier front at Appussaavik north of Nuussuaq/ Cape Walker in Qimusseriarsuaq/ Melville Bay. The hunters noted this is an area with polar bear maternity dens. Photo: E.W. Born

stable sea ice, (2) the occurrence of the North Water polynya (a large area of open water situated in the northern part of Baffin Bay and Ikeq/Smith Sound during the winter and well into May; Barber *et al.* 2001a, b) and (3) the settlement locations of the human population.

Typically, the northernmost part of the hunting area (Ikersuaq/Kane Basin) is covered with dense pack ice. Usually, the southernmost edge of this ice stretches from Anoritooq/ Kap Inglefield in Greenland to Pikiuliusarsuaq/Pim Island at Umimmaat Nunaat/Ellesmere Island in Canada from formation of new ice in the autumn until ice breakup next spring or summer. Fast ice covers the fjords in the central part of the Qaanaaq municipality, from Ullersuaq/Kap Alexander to Innaanganeq/Kap York, between October and the ice break-up, usually sometime in July (Teilmann *et al.* 1999; Born *et al.* 2004). This part of the area which forms the easternmost boundary of the North Water is characterised in the winter months by open-water and drift ice along the coast (Barber *et al.* 2001a, b). The land fast ice in Qimusseriarsuaq/Melville

Bay between Innaanganeq/Kap York and Kiatassuaq/Holm Island is present until relatively late in the spring. The drift ice of Baffin Bay is found to the west of this area. A lead of open water in the shear zone between the drift ice and the land fast ice stretches from Innaanganeq/Kap York to Kiatassuaq/Holm Island in the Upernavik municipality (Rasmussen 1919b, 1921; Siegstad *et al.* 2001). This lead widens when there is an easterly or northerly wind (Rasmussen 1921). The lead is called *"Qimusseriarsuup Oqaa"* – "The tongue of the Melville Bay" (B. Frederiksen, Nuussuaq, pers. comm. 2007); Rasmussen (1919b, 1921) refers to it as *"The mouth of the Sea"*.

Qimuseriarsuaq/Melville Bay is traditionally an important area for bear hunting (*e.g.* Rosing-Asvid & Born 1990) and is ideal polar bear habitat. The land fast ice provides an excellent breeding substrate for the bears' primary prey, the ringed seal, which are numerous in the region (Born *et al.* 1999). Large sections of the coast are intersected by glaciers, rendering some coastal areas inaccessible, and thereby providing seemingly ideal breeding areas for polar bears (Born 1995).

Conditions in the southernmost part of the Upernavik municipality are more affected by the relatively warm "West Greenland Current", which flows from the south. This means that there is "light" ice coverage or open water between the land fast ice and Baffin Bay's drift ice relatively early in the year (*e.g.* Buch 2001).

The polar night, where the sun never shows itself above the horizon, darkens the sky in the Qaanaaq municipality from late October or early November to around the middle of February, depending on the degree of latitude. In the Upernavik municipality, the polar night lasts from early or mid November to the end of January or early February, depending on the location (Berthelsen *et al.* 1990).

The distribution of polar bears and the bear catch is also affected by patterns of human habitation in the area. The settlements in the Qaanaaq municipality, which have a total population of 796 (2005), are grouped in the central part of the area which means that the hunting activity and traffic in this area is more intense than in Ikersuaq/Kane Basin and Qimusseriarsuaq/Melville Bay (the traditional bear hunting grounds). The population of the Upernavik municipality, which consisted of 2849 people in 2005, lives scattered along the coastline to the south of Qimusseriarsuaq (Fig. 2).

Hunting regulations and other factors that limit the hunt

This section contains a brief summary of the regulations governing polar bear hunting, which were in effect during the period of the interview survey (*cf.* Born & Rosing-Asvid 1989 and Rosing-Asvid & Born 1990 for further details concerning the development of regulations governing polar bear hunting in Greenland). General regulations governing the polar bear catch were introduced in Greenland on 1 January 1975. The rules have been slightly modified over time, but on the whole they have reserved the right to hunt polar bears for individuals who practice hunting as their primary occupation, have permanent residence in Greenland, and have strong ties to the Greenlandic society. The use of aeroplanes, helicopters and motorised vehicles (including snowmobiles), as well as boats larger than 40 BRT is forbidden when hunting polar bears or when travelling to and from hunting areas. Furthermore, the use of poison, traps, snares and semi- or fully-automatic rifles is also prohibited. All polar bears in Greenland are protected between 1 July and 31 August. However, since May 1988, the shooting of solitary adult males has been permitted year round. Cubs up to one year of age and their mothers are protected in all of Greenland. Moreover, there are regulations that prohibit the disturbance and uncovering of denning polar bears. New stricter hunting regulations were introduced on 1 January 2006, including the introduction of quotas on the Greenlandic polar bear catch and the full protection of females with cubs irrespective of their age (Anon. 2006, 2007). However, these new regulations have no relevance for our survey, which is based on data from polar bears caught before the end of 2005.

In 1967 Canada introduced quota limits on polar bear hunting including in the Northwest Territories (Urquhart & Schweinsburg 1984). Although hunters from Greenland were permitted to travel through Canadian territory along Ellesmere Island (Vibe 1968), they were *de facto* no longer permitted to catch polar bears on Canadian territory (*i.e.* to the west of the mid sector boundary in Ikeq/Smith Sound and Ikersuaq/Kane Basin); Kiliaan *et al.* (1978). The Melville Bay Nature Reserve was created in June 1980 (Anon. 1980). The reserve consists of an outer zone (I) where hunting is permitted and an inner zone (II), in which all hunting is prohibited. Those travelling between the Qaanaaq and Upernavik municipalities must also avoid zone II. It is however permitted to pursue a wounded polar bear into zone II, as long as the hunter can prove that the shot that wounded the bear was fired while the animal was outside of zone II.

Changes to the temperature and the sea ice cover

The air temperature in the Qaanaaq and Upernavik municipalities is continuously monitored. Since 1978, it has also been possible to obtain information from satellites about the sea ice and duration of ice coverage (*e.g.* Comiso & Parkinson 2004). In order to put the hunters' information on climate change into perspective, we provide a summary of the trends in air temperature and sea ice coverage based on these methods.

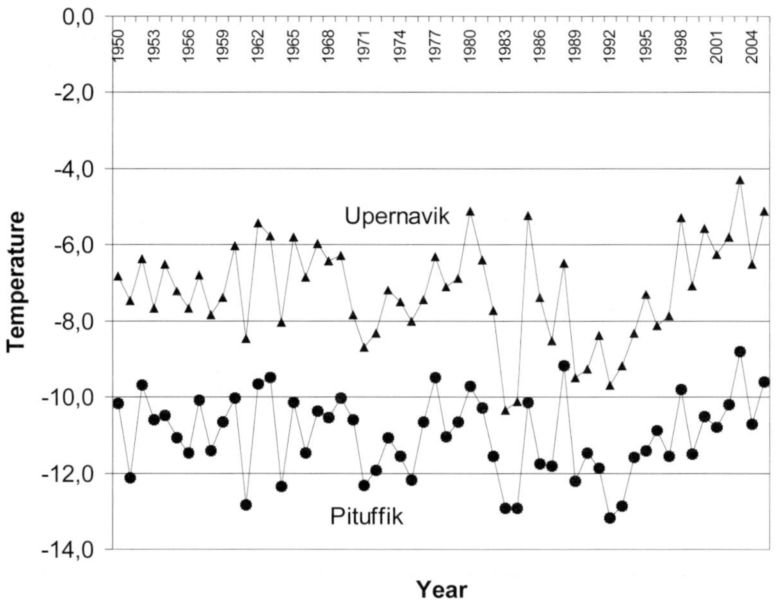

FIG. 3. The annual mean air temperature (°C) in the municipalities of Qaanaaq (measured in Pituffik/Thule Air Base) and Upernavik (measured in the town of Upernavik) between 1950 and 2005. Source: Cappelen 2006.

The annual mean air temperature in the Qaanaaq and Upernavik municipalities fluctuated widely during the period of time when polar bear catch data were available (1952–2005) (Fig. 3). From around 1950 until the early 1990s, the air temperature decreased in both municipalities. However, this decrease was not statistically significant (Qaanaaq: $r^2 = 0.039$, $z = -1.228$, $P = 0.219$, $n = 41$ years; Upernavik: $r^2 = 0.082$, $z = -1.811$, $P = 0.070$, $n = 41$). Between 1950 and 1989, the annual mean air temperature was -11°C in the Qaanaaq municipality (sd = 1.02, min.-max.:-12.9 and -9.2) and -7.3 °C in Upernavik municipality (sd = 1.24, min.-max.: -10.3 and -5.1°C) (Cappelen 2006). However, like the rest of Greenland, both areas have experienced a rapid and statistically significant increase in air temperature since the early 1990s (Fig. 3) (1990–2005, Qaanaaq: $r^2 = 0.596$, $z = 3.698$, $P = 0.0002$, $n = 16$ years; Upernavik: $r^2 = 0.785$, $z = 4.875$, $p < 0.0001$, $n = 16$). In the last 5 years of this study (2001–2005), the average yearly air temperature in the Qaanaaq municipality was -10.0 °C (sd = 1.83, min.-max.: -10.8 and -8.8 °C) and -5.6 °C in Upernavik (sd = 0.887, min.-max.: -6.5 and -4.3 °C) (Cappelen 2006).

An examination of the total annual extent of sea ice does not suggest any large changes to the North Water area (Fig. 4). However, the number of days with less than 50% ice coverage has increased significantly since 1996 (Fig. 5; $r^2 = 0.659$, $z = 2.997$, $P = 0.003$, $n = 10$ years), indicating that ice conditions have, on the whole, been reduced (*i.e.* less ice and more open water). In the eastern part of Baffin Bay, where the Upernavik municipality forms the boundary to the northeast, there has

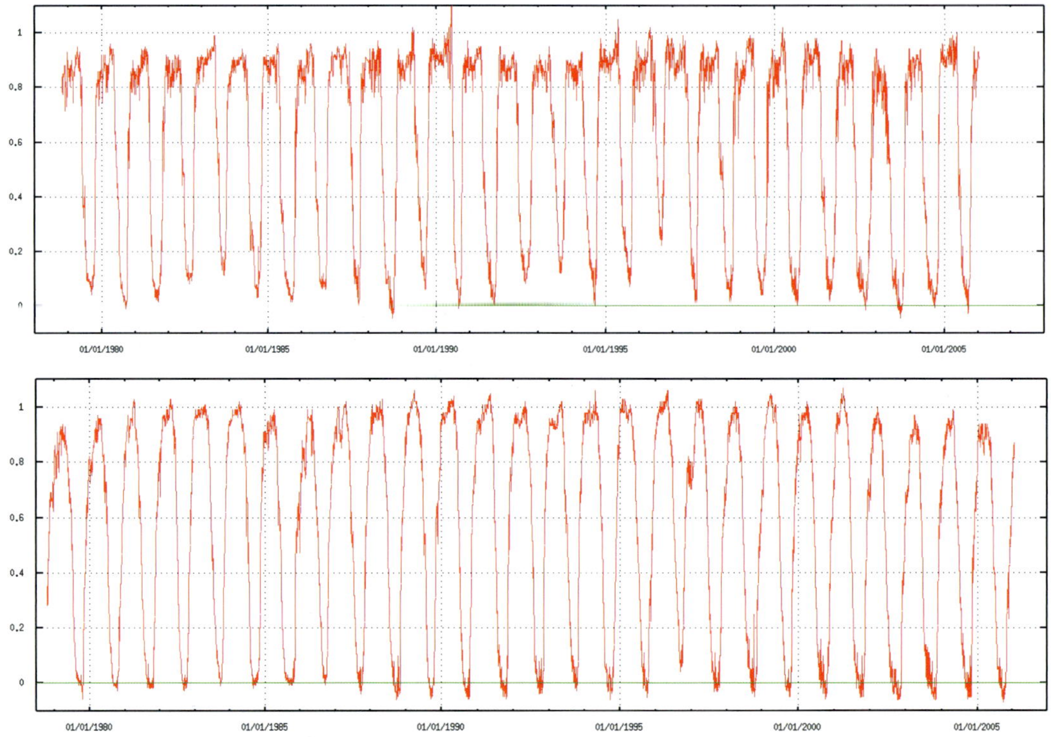

FIG. 4. The annual ice coverage in the North Water (*i.e.* where the Qaanaaq municipality is located; above) and the northeastern part of Baffin Bay *(i.e.* where Upernavik forms the eastern coasts; below) in the period 1979–2005. Each column indicates the relative amount of ice (height; scale from 0 to 1 where 0 = summer and 1= maximum ice coverage in the winter) and the duration of the ice coverage (width). Data recorded by satellites that register passive microwave radiation. Source: L. Toudal Pedersen, Danish Meteorological Institute, Copenhagen.

been a decrease in the total annual extent of sea ice between 2001–2005 (Fig. 4). In this area, the total number of days with less than 50% ice coverage has also increased markedly since the mid 1990s (Fig. 5; $r^2 = 0.493$, $z = 2.306$, $P = 0.02$, $n = 10$). Thus, both municipalities have experienced an increase in the air temperature and a decrease in sea ice coverage throughout the 1990s, with an acceleration in this process since the end of the 1990s.

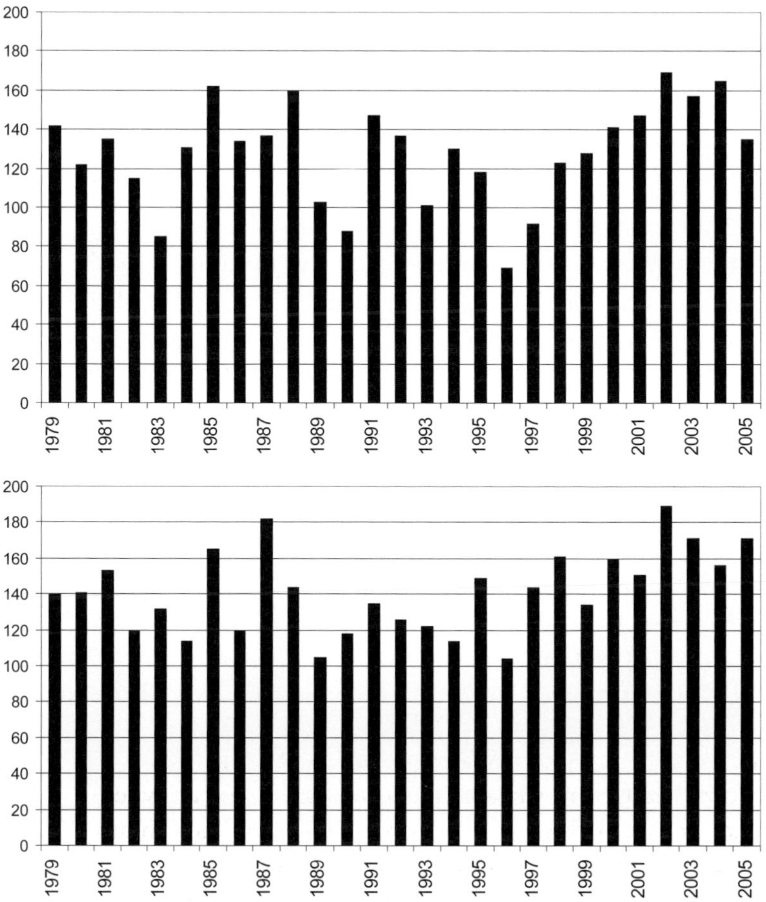

FIG. 5.
The number of days per year with less than 50% ice coverage in the North Water (*i.e.* where the Qaanaaq municipality is located; above) and the northeastern part of Baffin Bay (*i.e.* where Upernavik forms the eastern coasts; below) between 1979 and 2005. Data recorded by satellites that register passive microwave radiation. Source: L. Toudal Pedersen, Danish Meteorological Institute, Copenhagen.

Results

The interviewees

In both municipalities, the interviewees were a representative sample of experienced and active polar bear hunters who were largely responsible for the polar bear catch over the past five years. Informants in the Qaanaaq municipality tended to be a little older than those interviewed in the Upernavik municipality.

The Qaanaaq municipality

The locations where the 25 polar bear hunters were living at the time of the interviews are listed in Table 1. According to "*Piniarneq*" (DFHA, *in litt.* 2006), 52 hunters in the municipality had reported catching 179 polar bears in the preceding five year period (2001–2005; only the first nine months of 2005 are included). About 70% of these polar bears were reported by 22 of the 25 informants (three of the informants had not reported their catch of polar bears to "*Piniarneq*" during the period). The 25 interviewees represented ca. 38% of the hunters registered in the municipality at the start of 2006 (Table 1). However, as stated in "Materials, methods and background," the interviewees were selected on the basis of their knowledge and polar bear hunting experience.

The interviewees were aged between 32 and 69 years (mean = 49.5 years, sd = 9.4, n = 25), and the majority of them (n = 16) were between 40 and 49. They were between the ages of 12 and 30 when they shot their first bear (mean = 19.4 years, sd = 5.0; 64% were ≤19 years old); Fig. 6. In general, the informants in the Qaanaaq municipality had more years of experience of polar bear hunting than those in the Upernavik municipality.

The 25 hunters gave detailed information on an average of 13.5 polar bear catches (sd = 10.5, range: 1–39 polar bears / hunter); 15 of them described catching 11 or more polar bears. However, several of the hunters had caught, or participated in catching, many more bears than they could remember in detail.

Among the interviewees, 20 (80%) had shot a bear in the past 5 years (69 catches and a total of 93 animals) and they spoke of 17 more bear catches that had been made by 7 hunters who were not interviewed. Thus, the interviewees were predominantly highly experienced and currently active polar bear hunters.

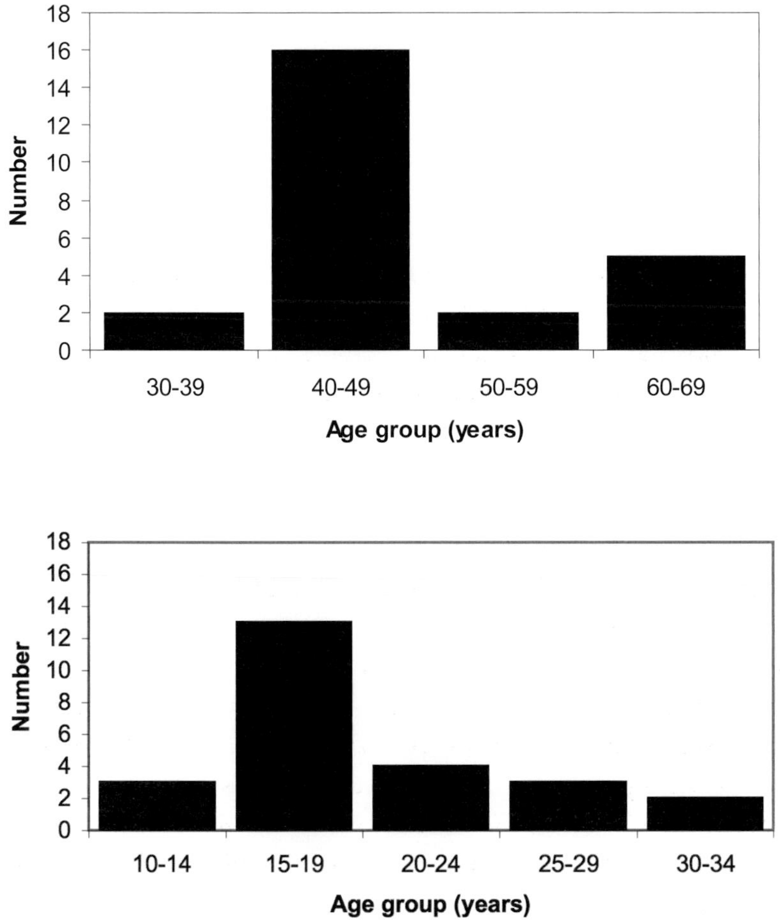

FIG. 6. The age distribution of the 25 polar bear hunters who were interviewed in the Qaanaaq municipality in February 2006 (above) and the age distribution when the hunters killed their first bear (below).

The Upernavik municipality

The distribution of the residences of the 47 hunters interviewed in the Upernavik municipality is shown in Table 1. According to "*Piniarneq*" (DFHA, *in litt.* 2006) 153 hunters in the municipality reported catching 444 polar bears during 2001–2005. Around 50% (n=220) of these bears were reported by 37 of the 47 interviewees (10 of the informants had not reported their catch of polar bears to "*Piniarneq*" in the period specified). The informants made up about 15% of all of the hunters registered in the municipality at the beginning of 2006 (Table 1), but as previously mentioned, they were specifically selected on the basis of their polar bear hunting experience.

Similar to the Qaanaaq municipality, the group of interviewees consisted mainly of experienced hunters in their 40s, but there was a wider range in their current ages, as well as in the age at which they had caught their first polar bear. The interviewees were between 18 and 75 years old (mean = 46.0 years, sd = 17.2, n

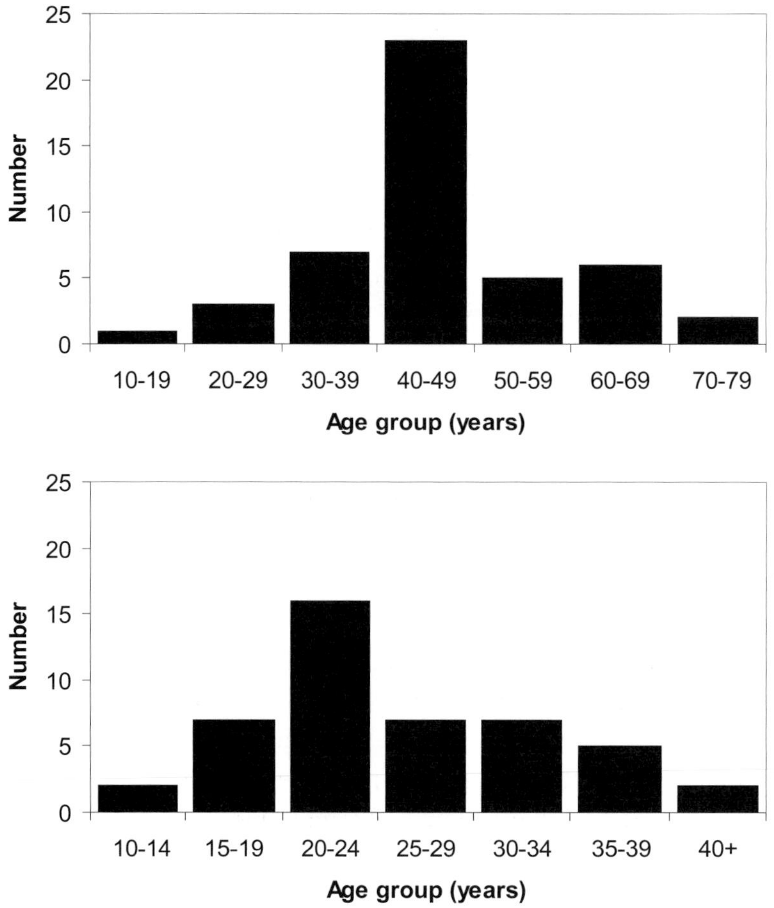

FIG. 7. The age distribution of the 47 polar bear hunters who were interviewed in the Upernavik municipality in February 2006 (above; n = 47) and their age distribution when the hunters killed their first bear (below; n = 46; the age of one hunter was not specified).

= 47) and about half of them (n = 23) were aged between 40 and 49. The age at which they had caught their first polar bear ranged from 12 to 51 years (mean = 25.6 years, sd = 8.4, n = 46); 19% were ≤19 years old when they had shot their first bear (Fig. 7). During the course of the interviews, the 47 hunters gave detailed reports of an average of 5.6 polar bear catches per hunter (sd = 5.4; variation: 1–23 polar bears/hunter). Eight of the hunters described 10 or more polar bear catches. In this municipality, several of the hunters had also caught many more bears than they could remember in detail. Around 66% (n = 31) had shot a bear in the past 5 years (a total of 111 catches and 143 animals in all) and in addition the catch of one bear by a hunter who was not interviewed was reported.

On the whole, the hunters in the Upernavik municipality had caught their first bear at a later age than those in the Qaanaaq municipality. It was perhaps of greater significance that the hunters in Upernavik's southern region (Area 4) had, on average, fewer years of bear hunting experience than the hunters in Area 3 (Nuussuaq-

Savissivik) and the area to the north (ANOVA and Fisher *post hoc* tests, $p<0.001$). In Area 4 there was an average of 16.5 years (sd = 10.9; n = 28) between the age of the hunter at the time of the interview and his age at the time of his first bear catch, while the difference in Area 3 was 27.5 years (sd = 12.9, n = 26) and in the area to the north of Savissivik it was 28.7 years (sd = 11.5, n = 17).

Hence, the hunters in the area to the south of Nuussuaq had less bear hunting experience (in terms of years) and furthermore had gained their experience mainly during the period of time when climate change was greatest.

Responses to the questions

The physical conditions and the sea ice coverage differ in the different hunting areas used by hunters from the settlements in the Qaanaaq municipality (see section on Physical conditions). We present the responses to questions in each sub-area from north to south (areas 1–3).

Hunters from the various settlements in the Upernavik municipality also utilise areas with different ecological conditions. Information from the municipality of Upernavik is therefore summarised from north to south, and Area 3 (Nuussuaq and Kullorsuaq) and Area 4 (Nutaarmiut to the town of Upernavik) are presented separately. The hunter's age is mentioned if it was considered to be relevant to his statement.

The Qaanaaq municipality

The responses to questions 1 and 2 ("How many bears do you think were caught in your settlement in 2005?" and "How many bears do you believe were caught in the municipality in total?") are summarised in the section entitled "Number of animals taken" (The catch in figures), p. 180.

Hunting and travelling conditions

3. Do you catch more bears than you used to?

Of the 17 hunters who were asked the question, 10 (59%) answered "yes", 5 (29%) answered "no" and the 2 remaining respondents (12%) did not have any specific opinion on the matter (Table 2).

A hunter from Siorapaluk explained that the catch had increased since they had started to travel by boat to the edge of the pack ice in the southern part of

	Total	Siorapaluk	Qaanaaq-Qeqertarsuaq	Savissivik
Yes	10	1	5	4
No/Same	5	1	2	2
No opinion	2	-	1	1
N_{total} (hunters asked)	17	2	8	7

Table 2. Distribution of the responses to question 3, from the Qaanaaq municipality: "Do you catch more bears than you used to?"

Ikersuaq/Kane Basin. The bears were caught while the hunters were out hunting Atlantic walruses (*Odobenus rosmarus rosmarus*) and narwhals (*Monodon monoceros*) in the area. Several hunters from Qaanaaq-Qeqertarsuaq reported that they had caught a greater number of bears, that the bears seemed to be easier to catch, and bears had come closer to inhabited areas and areas with regular human activity. One informant mentioned that it was no longer possible to leave meat caches without them being disturbed, even in the parts of Qaanaap Kangerlua/Ingelfield Bredning that are far inland.

In Savissivik, a number of interviewees indicated that the catch had increased due to the polar bears' closer proximity to land. Polar bears were caught on hunting trips where the intended prey was another animal (*e.g.* seals) or on trips to tend to seal nets. One hunter informed us that there had not been many polar bears when he moved to Savissivik (around 1975), and that at that time hunters had to travel far out onto the drift ice in Baffin Bay in order to hunt polar bears.

Five of the informants did not think that their own polar bear catch had increased in recent years.

4. Are more bears caught here in the settlement than previously?

Of the 23 hunters who were asked this question, 21 (92%) answered "yes", 1 (4%) answered "no" and the remaining hunter had no particular opinion on the matter (Table 3).

Several of the more detailed explanations coincided with the responses given for the previous question. One 65-year-old hunter from Siorapaluk, who had previous experience from the northern region as well as from Umimaat Nunaat/Ellesmere Island in Canada, said that the catch of polar bears in the "local" area had indeed increased. Furthermore, he explained that polar bears had nearly been "exterminated" in the period between the 1950s and 1970s, but had increased in numbers since the start of the 1980s. Another hunter from Siorapaluk believed that

	Total	Siorapaluk	Qaanaaq-Qeqertarsuaq	Savissivik
Yes	21	6	9	6
No/Same	1	-	-	1
No opinion	1	-	1	-
N_{total} (hunters asked)	23	6	10	7

Table 3. Distribution of the responses to question 4, from the Qaanaaq municipality: "Are more bears caught here in the settlement than previously?"

the number of bears caught by individuals from his settlement was unchanged and that it varied in any case from year to year. He had, however, heard that the catch taken by hunters from the town of Qaanaaq had increased in recent years. A third informant from the same settlement said that there seemed to be a greater number of polar bears than in earlier times.

The hunters in Qaanaaq town and Qeqertarsuaq stated that it had become easier to catch polar bears, and two hunters said that they had ceased making long bear hunting trips due to the bears' increased occurrence in the "local" area. One 46-year-old informant said that a greater number of bears were caught now compared to when he was young, while a 54-year-old man recalled that when he was a child the bear hunters who travelled north to Nuussuaq/Washington Land in the Ikersuaq/Kane Basin region often used to come home empty handed. Since 1975 when he began to hunt, the bear hunts had generally been successful. This particular hunter believed that the changes in conditions were partly due to the ban on hunting female bears with cubs <1 year old (from 1 January 1975 in Greenland; author's note) and partly due to the fact that the bears migrate from Canada where bear hunting restrictions also apply. A 69-year-old hunter thought that the number of bears had increased in recent years due to the ban on hunting females with cubs <1 year old. He believed that bears had also generally become less timid.

Two hunters from Savissivik, aged 48 and 63 respectively, responded to this question by saying that more polar bears came closer to the coast because of the decrease in sea ice.

One 46-year-old hunter from the same settlement, who said that the catch had not increased and "appeared to be the same," added that he had not lived in the settlement for very long. Another informant did not answer the question directly but said that hunters saw a lot of polar bears when they travelled to the north because the regions to the north (*i.e.* Ikersuaq/Kane Basin) are covered in sea ice.

5. Do you use regular routes when hunting polar bears?

Of the 10 hunters who were asked this question, 9 (90%) answered "yes," while the remaining hunter did not have any opinion on the subject (Table 4).

When travelling in the spring to the northern part of Ikeq/Smith Sound and Ikersuaq/Kane Basin by skiff, the hunters' route follows the coastline north via Ullersuaq/Kap Alexander to the southern edge of the Ikersuaq/Kane Basin pack ice, which is situated between Anoritooq/Kap Inglefield in Greenland and Pikiuliusarsuaq/Pim Island in Canada. However, when travelling to the north on a dogsled in the late winter, it is possible to travel via the Inland Ice Cap in Nunarsuaq/Inglefield Land. The routes which pass over the ice cap are shortcuts and are safer and more passable than the sled route to the west around Ullersuaq/Kap Alexander

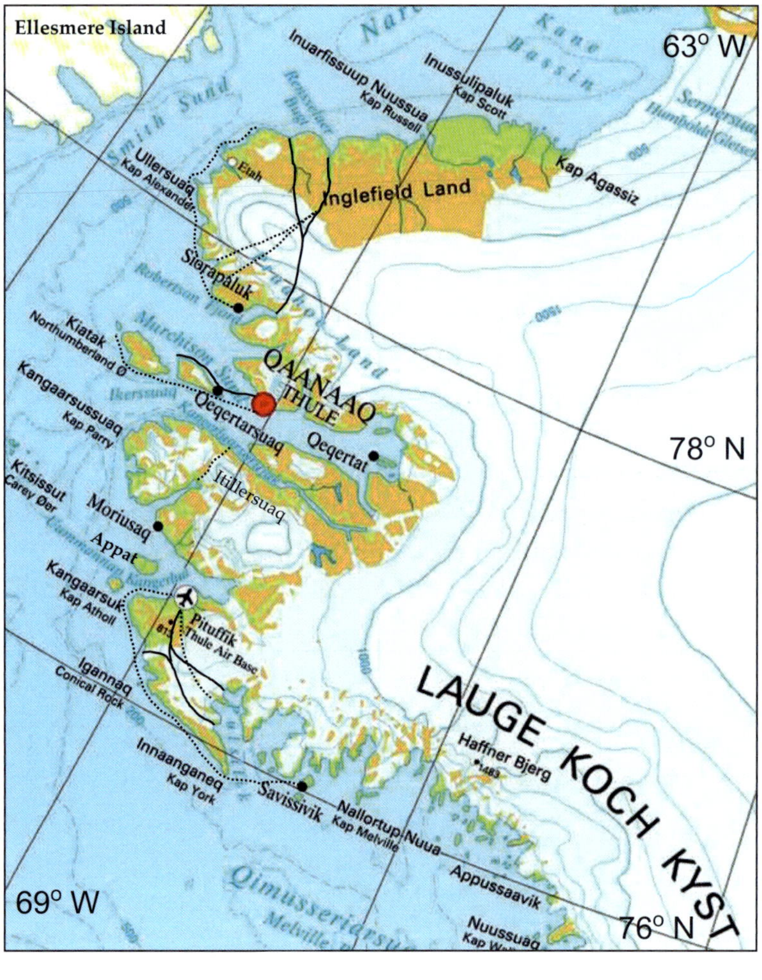

FIG. 8. Previously used (dotted line) and currently used (solid line) sled routes as drawn by seven hunters during the interviews in the Qaanaaq municipality in February 2006.

	Total	Siorapaluk	Qaanaaq-Qeqertarsuaq	Savissivik
Yes	9	4	4	1
No	0	-	-	-
No opinion	1	-	1	-
N_{total} (hunters asked)	10	4	5	1

Table 4. Distribution of the responses to question 5, from the Qaanaaq municipality: "Do you use regular routes when hunting polar bears?"

or over the glacier just behind this promontory. Another route goes via the glacier at Neqi to Nunarsuaq/Inglefield Land.

One hunter reported that bear hunters often go to Kiatak/Northumberland Island, while another informant who lives in Savissivik stated that it was usual to travel north along the outer coast to the north of Innaanganeq/Kap York or via the ice cap between Puisilivik (Puisilluusarsuaq/De Dødes Fjord) and Pituffik/Thule Air Base at Parker Snow Bay (Fig. 8). However, the routes in the Qaanaaq municipality have changed because of the glaciers melting (see section "Have your routes changed in recent years?", p. 37).

One informant pointed out that the conditions of the sea ice generally determine the choice of route.

6. Do a greater number of polar bears come to visit/come of their own accord?

Bears which come to inhabited areas of their own accord are called "*tikeraat*". Nineteen (86%) of the 22 respondents replied by saying "yes", while 2 (9%) responded by saying "no" and one had no opinion on the matter (Table 5).

Two hunters from Siorapaluk (aged 43 and 44) were of the opinion that bears had come more often and closer to inhabited areas and areas with regular human activity

	Total	Siorapaluk	Qaanaaq-Qeqertarsuaq	Savissivik
Yes	19	2	9	8
No/Same	2	2	-	-
No opinion	1	1	-	-
N_{total} (hunters asked)	22	5	9	8

Table 5. Distribution of the responses to question 6, from the Qaanaaq municipality: "Do a greater number of polar bears come to visit/come of their own accord?"

since the 1990s (it was unclear whether they meant the start of the 1990s). However, two other hunters (aged 58 and 64) from the same settlement said that polar bear visits in the settlement were very uncommon (1 bear in 1997 and 1 in 2005 were mentioned by these informants). Another hunter (aged 46) also from Siorapaluk, who otherwise expressed no particular opinion in response to this question, had only heard of a single case where a polar bear had wandered into the settlement.

A number of informants from Qaanaaq-Qeqertarsuaq expressed the opinion that polar bears are now seen more frequently in and around inhabited areas. One 49-year-old hunter stated that this used to be a seldom occurrence but that it had become more common since the 1980s. Before that time, when he was a child (*i.e.* in the 1960s), the bear hunt had taken place on Umimmaat Nunaat/Ellesmere Island in Canada and around Savissivik in northern Qimusseriarsuaq/Melville Bay. A 54-year-old hunter explained that in the 1970s no bear tracks were observed at Sanerarsuaq (the coastal area between Innaanganeq/Kap York and Kangaarsuk/Kap Atholl), but that it was now quite common to see the tracks of several bears when travelling along that stretch of coastline. Some of the informants stated that polar bears now come closer to Qaanaaq (one visit to the beach was mentioned) and described the presence of tracks going into and out of Qaanaap Kangelua/Inglefield Bredning. Furthermore it was indicated that bears now occur closer to Siorapaluk, Kiatak/Northumberland Island and Moriusaq.

The hunters from Savissivik also reported that polar bears have begun to appear in closer proximity to land in recent years; especially in the area around the settlement where once nine bears had been killed (the year in which this occurred was not specified). A 40-year-old informant reported six polar bears had come close to the Savissivik area and were shot in December 2002. Another informant's (aged 32) response included the observation that the polar bears had become increasingly lean.

Several explanations were offered for the bears' presence in the proximity of inhabited areas: (1) Polar bears tend to be more widely distributed when they are more abundant and they also prefer to be close to each other when they move around (*i.e.* polar bears have a tendency to keep track of each other), (2) there are more bears and they are hungry and so they come closer to settlements, (3) the lean bears come from offshore areas in order to find food and they get in the way of each other's hunting activity, (4) they come closer to the areas where there is ice, because the sea "out there" no longer freezes and has unfavourable ice conditions, and (5) they come closer when the sea ice begins to form in the autumn and when there is a new moon.

Only a small number of the reported bear catches were characterised as bears that had wandered of their own accord into inhabited areas (see section "Hunting methods" in "The catch in figures" p. 187).

7. Have your routes changed in recent years?

Sixteen informants were asked this question and 14 of them (ca. 88%) answered "yes", 1 (ca. 6%) answered "no" and one informant did not express any particular opinion on the subject (Table 6).

	Total	Siorapaluk	Qaanaaq-Qeqertarsuaq	Savissivik
Yes	14	4	6	4
No / Same	1	-	1	-
No opinion	1	-	1	-
N_{total} (hunters asked)	16	4	8	4

Table 6. Distribution of the responses to question 7, from the Qaanaaq municipality: "Have your routes changed in recent years?"

Hunters from all of the areas in the municipality had experienced a change in the hunting routes. Thirteen of them said that the routes had changed during recent years because the sea ice forms later, is less stable, and lasts for a shorter period of time. Those who hunted in Ikeq/Smith Sound – Ikersuaq/Kane Basin stated that it is no longer possible to drive a dogsled to the north along the coast due to a lack of sea ice around Ullersuaq/Kap Alexander and because there is no longer a passable "ice-shelf"

Nunataami/Cape Jackson on the southern coast of Washington Land (April 1994). Before the sea ice decreased in Ikersuaq/Kane Basin polar bear hunters from Qaanaaq municipality reached this far north by dog sled. Photo: E.W. Born to drive over (ice-shelf, *"qaanngoq"*, *i.e.* a shelf of sea ice attached to the coast which can be up to several metres wide and which is formed during the changing of the tide). Due to the lack of sea ice, hunters travel via sled routes over the glaciers but these have also undergone changes in recent years. As a result of the glacier ice melting it is no longer possible to drive northwards via the Itillersuaq glacier at Pitoraarfik/Kap Chalon or via the glacier at Neqi. The only passable glaciers are Itilliarsuq (the glacier east of Ullersuaq/Kap Alexander) and Arfalluarvik/Diebitsch Glacier. One route passes over the Ikineq glacier at the head of the Siorapaluup Kangerlua/Robertson Fjord (Fig. 8). One of the hunters had heard from other people that it may also be possible to drive over the glacier by Pitoraarfik/Kap Chalon (Clement Markham Glacier).

In the central area (*i.e.* Area 2) areas of open water and unsafe ice occur frequently between Qeqertarsuaq/Herbert Island and Kiatak/Northumberland Island. Due to the unsafe ice and holes in the ice (*aakkarneq*) it is often impossible to take a dog team along the southern side of the island of Qeqertarsuaq in a westerly direction towards the ice edge in the Kiatak area. Hunters must therefore take a route which runs along the northern side of Qeqertarsuaq.

The well known and much used route over Itillersuaq/Politikens Bræ is no longer passable due to large crevasses in the front of the glacier by Ikeq/Hval Sund (see also the section "Have you seen changes to the glaciers?", p. 61); Fig. 8. In recent years there have been areas of unstable ice and open water between Kangaarsussuaq/Kap Parry and Appat/Saunders Island, and the hunters have therefore been forced to stay close to land when travelling to hunt walruses in the area around the westerly point of Appat.

Hunters from Savissivik, who mostly hunt in Qimusseriarsuq/Melville Bay and on the offshore drift ice to the west in Baffin Bay, reported that the routes had changed due to a decrease in ice coverage and because the ice is unsafe. For example, the ice around Innaanganeq/Kap York has become unstable, whereas it had previously been possible to go hunting by dogsled on the ice in this area in November. In recent years it has only been possible from January onwards. The lack of ice and/or unsafe ice in this area – and in the area along Sanerarsuaq (the coast between Innaanganeq and Kangaarsuk/Kap Atholl) – have made it more difficult to drive a dogsled along the coastal route to the north towards Pituffik. Furthermore, it is no longer possible to drive a dogsled to Igannaq/Conical Rock which is situated ca. 4 km to the west of Sanerarsuaq. One informant mentioned that it is not even possible to drive to the point of Serfarmiut anymore, because there is a hole in the ice. Moreover, the crevasses in the glacier mean that the dogsled route between Illaarsuk/Sidebriksfjord and Ikera/Meteorit Bugt is impassable. Other glaciers are now used as routes when travelling between Savissivik and Pituffik (Fig. 8; see also the section "Have you seen changes to the glaciers?", p. 61).

A 32-year-old hunter mentioned how he used to set out seal nets just to the east of Nallortuup nuua/Kap Melville but that the occurrence of unsafe ice and holes in the ice in recent years had prompted him to place his nets further to the west by Sermipaluk to the west of Akuliaruseq.

It is no longer possible to take a dog team onto the offshore drift ice in order to hunt polar bears because the ice is so thin that it has become too unsafe. A 32-year-old hunter recounted that the last time he had been bear hunting on the offshore ice was in 1994, and that since then driving a dogsled there had become increasingly problematic. A 48-year-old hunter, who articulated that it was no longer possible to drive a dogsled "offshore", mentioned that this used to be the only way to encounter bears in the past. He added that nowadays, since the ice no longer forms properly, the polar bears come closer to land and that one can encounter them while setting seal nets. The same hunter spoke of how hunters have begun to catch polar bears all around the Savissivik area and that they now take their dog teams to the base of the fjord, by Innaanganeq/Kap York (Puisilluusarsuaq/De Dødes Fjord), in order to hunt bears. The base of the fjord is the only place which is covered with ice and this ice attracts polar bears.

A 43-year-old hunter from Siorapaluk responded to the question by saying that the routes had changed in the sense that it is no longer necessary to travel north to Ikersuaq/Kane Basin because bears can now be encountered when travelling by boat north in Ikeq/Smith Sound.

However, a 49-year-old hunter from Qaanaaq, who reported catching a total of four polar bears, was of the opinion that the routes of travel had remained unchanged.

8. *Are a greater number of polar bears caught from boats now than in the past?*

Of the 22 informants who were asked this question, 13 (59%) answered "yes", 8 (36%) answered "no" and 1 (5%) had no opinion on the subject (Table 7).

The hunters living in Siorapaluk spoke of how hunting patterns have changed as a result of changes to the ice in Ikeq/Smith Sound and Ikersuaq/Kane Basin. For example, one 44-year-old hunter recounted that there had been a great deal of "old" ice (in this

	Total	Siorapaluk	Qaanaaq-Qeqertarsuaq	Savissivik
Yes	13	3	6	4
No/Same	8	1	4	3
No opinion	1	-	1	-
N_{total} (hunters asked)	22	4	11	7

Table 7. Distribution of the responses to question 8, from the Qaanaaq municipality: "Are a greater number of polar bears caught from boats now than in the past?"

In recent years a larger fraction of polar bears have been shot from cutters and skiffs in the northern parts of Ikeq/Smith Sound. Photo: J. Danielsen

case ice that formed in the fall and remained until ice break-up the next spring – author's note) in Ikersuaq in the 1970s and 1980s but that since around 1996, the ice had formed later, been thinner and broken up earlier. These changes have resulted in an earlier start to the boat-based hunting season. Another informant (aged 58) told the interviewer that it had not been possible to travel by boat in the northernmost part of Ikeq in the past because of the ice, while another hunter (aged 46) said that a number of bears had been caught by boat at the southern edge of the pack ice in Kane Basin since 2003.

Six hunters from Qaanaaq-Qeqertarsuaq also indicated that the number of polar bears killed from boats had increased. According to two informants (aged 44 and 49 respectively), an increased number of polar bears have been sighted during the open water season and polar bears are easier to find because they tend to come closer to land. Moreover, polar bears have been observed swimming in the water around Moriusaq. However, a 54-year-old hunter believed that one reason for the increase in bear sightings is the increase in boating traffic. When he was young, there were only three cutters in Qaanaaq. The relationship between the increase in the use of cutters and skiffs and the increase in the number of polar bear sightings was also mentioned by another hunter (aged 46) who thought that there may also be a relationship between the increased speed and range of travel, made possible by the use of skiffs, and the perceived increased occurrence of polar bears.

Several informants from Savissivik were also of the opinion that a greater number of bears are caught from boats than in earlier times. A 48-year-old hunter said that this increase is due to the introduction of skiffs with outboard motors, which makes it possible to travel more quickly than in the past. He also mentioned there had only been a small number of cutters in the past, from which few bears had been shot.

Three of the hunters who answered "yes" to this question did however add that bears that are seen from the boats (*i.e.* usually in the summer) are not usually shot as their fur is in poor condition because the animals moult at that time of year.

Five of the eight informants who answered "no" to this question also cited the poor condition of the fur during the boating season as the reason why there has been no change in the number of bears shot from boats. A 47-year-old hunter from Siorapaluk recounted how he had always taken part in polar bear hunts from a boat, along the ice edge between Anoritooq/Kap Inglefield and Pikiuliusarsuaq/Pim Island, and that there had been no change to the occurrence of polar bears there. A hunter from Savissivik mentioned that one or two bears are taken every summer for their meat, while another informant explained that when polar bears are occasionally shot from boats in the Savissivik area, primarily adult bears are killed.

From the information provided about the individual bear catches, it emerged that there has indeed been an increase in the proportion of bears caught from boats in the municipality of Qaanaaq (see section "Hunting methods" under "The catch in figures", p. 187).

9. Are a greater number of polar bears caught on land than in the past?

Ten informants were asked this question; 2 of them (20%) answered "yes" and the remaining 8 answered "no" (Table 8). In general, the answers to this question were uninformative about trends indicating an increase or decrease in frequency of bears on land. On the whole, the responses implied that it is unusual to encounter polar bears on land.

	Total	Siorapaluk	Qaanaaq-Qeqertarsuaq	Savissivik
Yes	2	1	1	-
No/Same	8	1	6	1
No opinion	-	-	-	-
N_{total} (hunters asked)	10	2	7	1

Table 8. Distribution of the responses to question 9, from the Qaanaaq municipality: "Are a greater number of polar bears caught on land than in the past?"

One of the informants who responded by saying "yes" to this question said that some polar bears had been shot on land at Kiatak/Northumberland Island in recent years, and bears are often encountered in that area during the summer and autumn.

The information provided by the informants who answered "no" or "unchanged" was not specific. One of the respondents reported that some people had observed polar bears on the shore at Kiatak in the spring, adding that some hunters catch polar bears on land and that bears on occasion move inland to flee hunters. However, he also said that he had never heard of polar bears remaining on land for longer periods of time. Another informant mentioned that he had only heard of polar bears being caught on land at Kiatak when people had placed meat depots on the island. He added that the bears come closer because they seek food and can smell the food on land. Other informants responded by saying that they had never seen or heard of bears coming onto the shore. One hunter said that the term "bears on land" only applied to "visiting" bears while another informant recalled having shot a bear on land himself. See also section "Hunting methods" in "The catch in figures", p. 187.

Occurrence of polar bears and dens

10. Do the bears occur in particular areas?

All of the 19 informants who were asked this question answered in the affirmative (Siorapaluk: 4; Qaanaaq-Qeqertarsuaq: 9; Savissivik: 6).

General statements made in response to this question were (1) that the occurrence of bears varies depending on the season, but nowadays they come closer to the coast when there is dense ice coverage and (2) that bears are often encountered on the ice edge, in areas where there are icebergs and in ice-covered areas. A hunter form Siorapaluk said that the bears flee from Canada and that they occur more and more frequently in the Qaanaaq municipality.

Some areas were identified as specific "bear areas" and of these the following were pointed out as being winter and spring "bear areas" by the hunters from Siorapaluk: In the Ikersuaq/Kane Basin primarily Nuussuaq/Washington Land, Sermersuaq/Humboldt Glacier, the north coast of Nunarsuaq/Inglefield Land (for example Inuarfissuaq/Marshall Bay) and the southern edge of the Kane Basin pack ice between Anoritooq/Kap Inglefield and Pikiuliusarsuaq/Pim Island. Washington Land and Umimaat Nunaat/Ellesmere Island had reportedly been traditional "bear areas" in the past, but some informants stated that polar bears now occur more frequently along the northern coast of Nunarsuaq/Inglefield Land, in Ikeq/Smith Sound, along the south coast of Kiatak/Northumberland Island, and in the

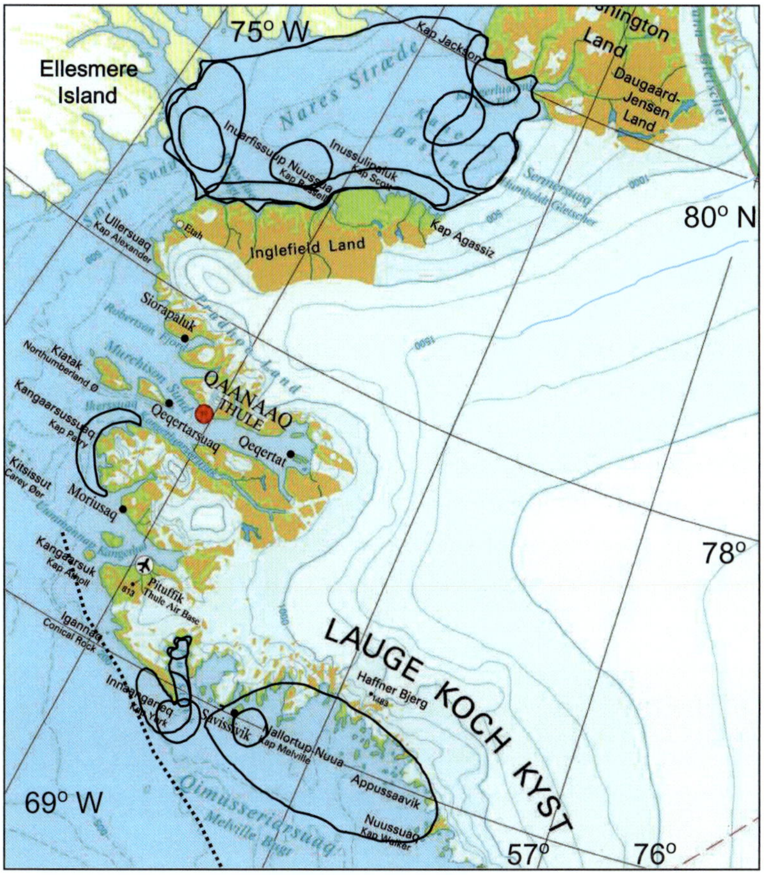

FIG. 9.
The general winter and spring distribution of the polar bears, as drawn on a map by nine hunters during the interviews in the Qaanaaq municipality in February 2006. The dotted line indicates the zone between the fast ice and the pack ice, which was mentioned as an important polar bear area by one hunter.

Innaanganeq/Kap York-Qimusseriarsuaq/Melville Bay area. The offshore drift ice in the Upernavik municipality was also identified as a "bear area" (Fig. 9).

The hunters from Qaanaaq-Qeqertarsuaq identified the following areas as "bear areas": Kiatak and the drift ice west of this island, Qaanaap Kangerlua/Inglefield Bredning, the area around Moriusaq, and the northern part of Qimusseriarsuaq/Melville Bay. Some informants mentioned that bears come in closer to the shore in November – even as far east as the Qaanaaq town area, and that Kiatak is also a "bear area" in the autumn.

In Savissivik, the informants also identified specific areas where polar bears tend to occur: the Innaanganeq/Kap York area, Puisilluusarsuaq/De Dødes Fjord ("the area where the glacier breaks off into the fjord and produces icebergs is a favourite place"), the area just south of Salleq/Bushnan Island and the area around Nallortuup nuaa/Kap Melville where there are stranded icebergs (Fig. 9).

The drift ice in Ikeq/Smith Sound, the area around the westerly point of Kiatak and the areas with glaciers in general were identified as the polar bears' summer ranges. (Fig. 13; see also the section "Where do the bears spend the summer?", p. 49).

11. Where have you seen bear tracks?

The purpose of this question was to gain a better insight into the occurrence of polar bears in general. All of the 11 informants (Siorapaluk: 5; Qaanaaq-Qeqertarsuaq: 6) that were asked this question were able to identify areas where they had seen bear tracks. Furthermore, some of the hunters drew on a map to indicate the areas where bear tracks are seen regularly (Fig. 10).

One hunter was of the opinion that at present, bear tracks are most often seen on land and on the ice, while another said that since the 1990s and into the 2000s bear tracks have been observed all over. Some of the informants said that there are a lot of bear tracks along the front of Sermersuaq/Humboldt Glacier and that they had

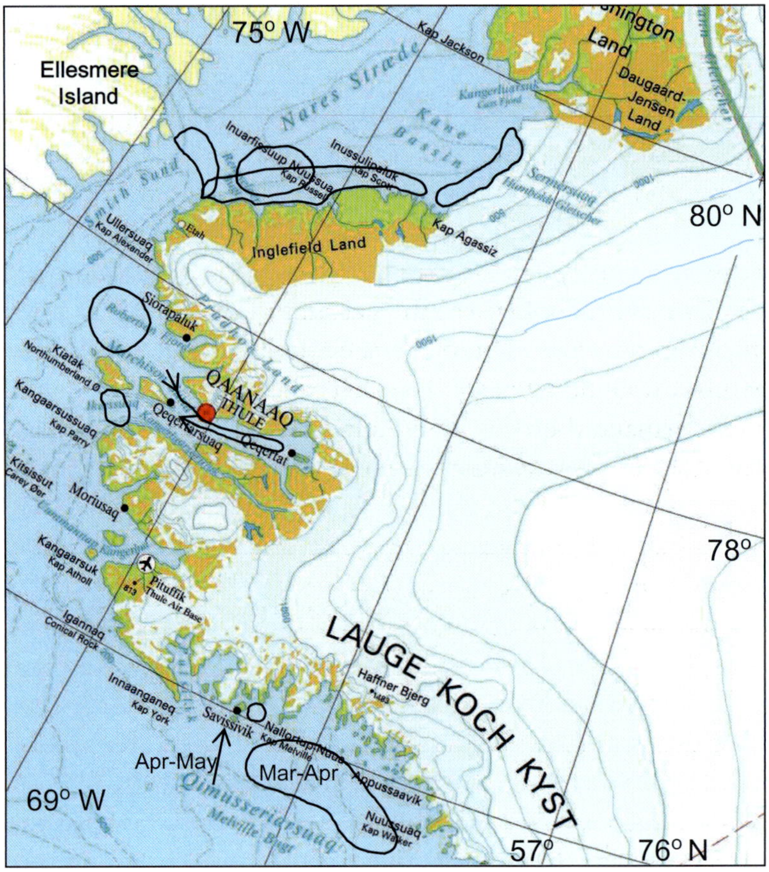

FIG. 10. Areas where bear tracks are often seen drawn by eight hunters during the interviews in the Qaanaaq municipality in February 2006.

recently begun to observe many tracks around Nunarsuaq/Inglefield Land, whereas in the past tracks had mostly been spotted further north at Nuussuaq/Washington Land. The ice edge between Anoritooq/Kap Inglefield and Pikiuliusarsuaq/Pim Island was also identified as an area where many bear tracks are observed in the springtime. According to one informant, the tracks in this area are mainly of adult bears. Another interviewee explained that the tracks in that area come from both directions (*i.e.* from Canada to Greenland and vice versa). A number of tracks had also been observed around the mouth of Iluleerloq/Murchison Sound and in Qimusseriarsuaq/Melville Bay. In the latter area, the tracks were said to indicate a southerly direction of travel in April (against the prevailing southerly wind) towards the scent of seals pups (see also the responses to question 11 from Area 3 in Upernavik, p. 100).

12. Do the bears have regular migration routes?

This question was asked to 9 interviewees. Eight (89%) answered "yes" and 1 (11%) answered "no" (Table 9). The interviewees made general statements which indicated that polar bears (1) migrate often in the mating season and frequently choose routes which run along the front of glaciers (*e.g.* along the front of Sermersuaq/Humboldt Glacier), (2) the bears move towards areas where the ice has broken up and is in motion at the end of March and in April, (3) the bears' movement patterns depend on the ice conditions, but also by their preference for being in close proximity to each other.

Furthermore, the informants spoke of some regular routes or migration patterns. A hunter from Qaanaaq identified a route that runs in a southerly direction at Nuussuaq/Washington Land which is used in the springtime and where the bears may even sometimes trample a broad, densely packed "road" over the ice. According to the same informant, there is a migration of bears in the spring from Sermersuaq/Humboldt Glacier in a westerly direction towards Bache Peninsula

	Total	Siorapaluk	Qaanaaq-Qeqertarsuaq	Savissivik
Yes	8	1	5	2
No	1	-	-	1
No opinion	-	-	-	-
N_{total} (hunters asked)	9	1	5	3

Table 9. Distribution of the responses to question 12, from the Qaanaaq municipality: "Do the bears have regular migration routes?"

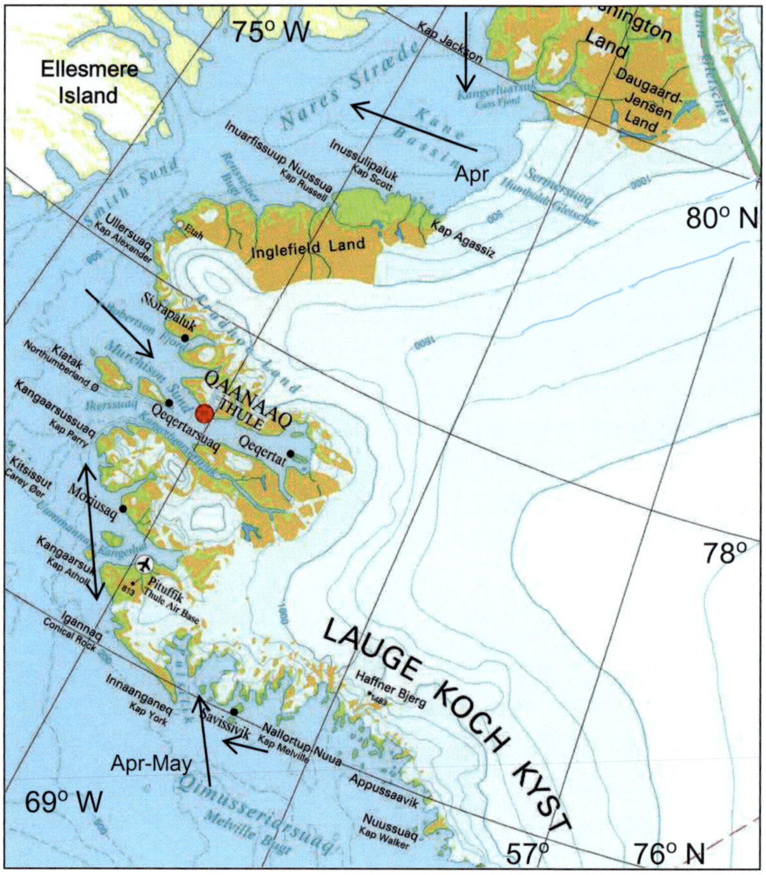

FIG. 11. Polar bear migration routes drawn by four hunters during the interviews in the Qaanaaq municipality in February 2006.

at Umimaat Nunaat/Ellesmere Island (Fig. 11). One interviewee mentioned that many bear tracks come from a westerly direction, in the mouth of Iluleerloq/ Murchison Sound, as well as further into the fjord at Qaanaap Kangerlua/Inglefield Bredning. Another informant said that the bears travel in both directions along the coast around Kangaarsussuaq/Kap Parry, while another was of the opinion that the bears mostly travel in a northerly direction in that area. In the Savissivik area, a hunter spoke of a migration which takes place in April–May towards the ringed seal pups at Illaarsuk/Sidebriks Fjord, while another reported a westerly migration of polar bears in the area around Salleq/Bushnan Island (Fig. 11).

13. Have you seen very small bear tracks (i.e. from cubs of the year)?

The purpose of this question was to ascertain the extent to which polar bears are breeding in Greenland. Observations made in the springtime of family groups with very young cubs (*"ateqqaaq/-t"*; "cubs of the year" that are born during the winter and have left the den with their mother) or of the tracks left by such groups, might imply that the cubs were born in a den in Greenland.

Of the 25 respondents, 24 (96%) answered "yes" to this question, while 1 respondent answered "no" (Table 10). Several areas were identified as locations where the tracks of very young cubs had been observed (Fig. 12).

Eighteen of the respondents drew pictures in order to indicate the approximate size of the tracks or "footprints" made by the small cubs. The average width of the tracks that were drawn was around 6.6 cm (sd = 1.5; min.-max.: 4.5-9.0 cm, n = 18) and the average length was ca. 7.4 cm (sd = 1.9; min.-max.: 4.5-10.0 cm, n = 18). The tracks of small bears were often described as being "the same size as dog tracks", which confirms that the cubs were 0-year-olds that had come out of the den in the spring of the same year.

Such tracks had often been observed along the coastline of Nunarsuaq/Inglefield Land and all along Sermersuaq/Humboldt Glacier, and are now more frequently observed than in earlier times – even close to Qaanaaq town. The

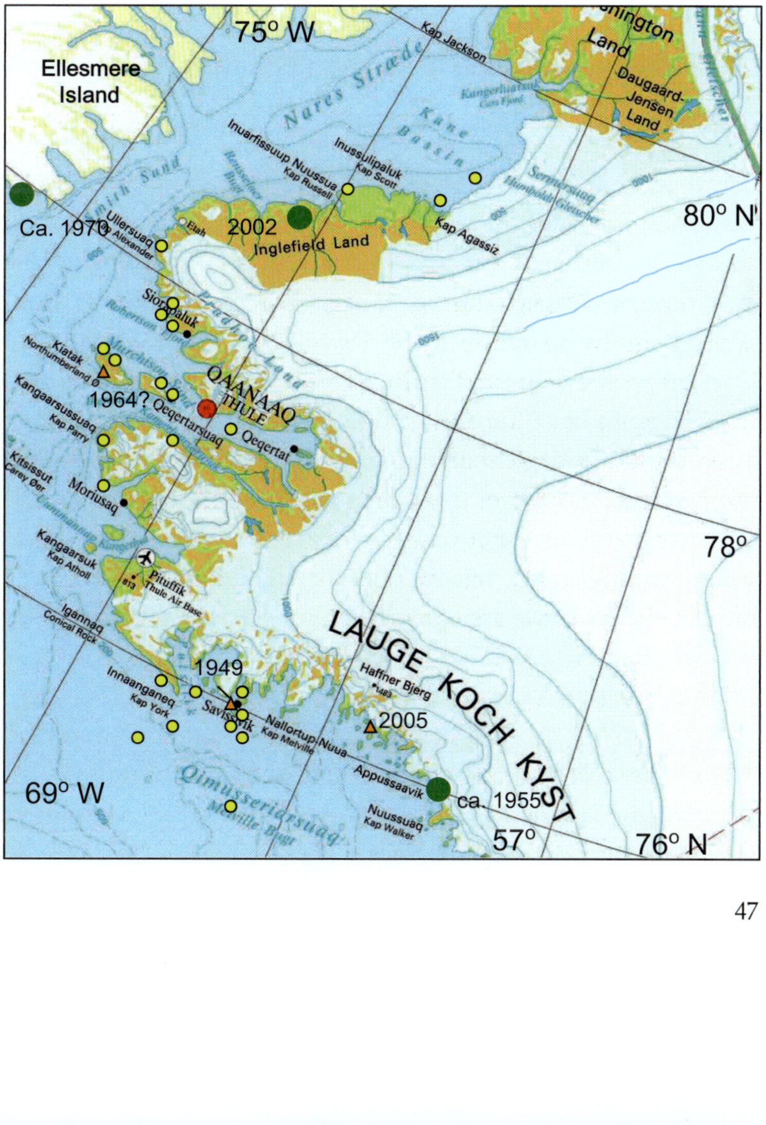

FIG. 12. Observations of tracks made by female polar bears with newborn cubs and dens, as drawn on a map by 17 hunters during the interviews in the Qaanaaq municipality in February 2006. Yellow dots = tracks; orange triangles = temporary dens; green dots = maternity dens.

	Total	Siora-paluk	Qaanaaq-Qeqertarsuaq	Savissivik
Yes	24	6	11	7
No	1	-	-	1
No opinion	-	-	-	-
N_{total} (hunters asked)	25	6	11	8

Table 10. Distribution of the responses to question 13, from the Qaanaaq municipality: "Have you seen very small bear tracks?"

respondents from the northernmost areas talked about observations of small tracks in the Nuussuaq/Washington Land area, in front of Sermersuaq/Humboldt Glacier and close to Qaqqaitsut/Kap Agassiz on Nunarsuaq/Inglefield Land. In the central part of the Qaanaaq municipality, these kinds of tracks have been observed to the west of Siorapaluk and in Siorapaluup Kangerlua/Robertson Fjord, around Kiatak/Northumberland Island (spring 2005), at Kuugarsuaq on the southern part of Qeqertarsuaq/Herbert Island (spring 2005) at Qalluusat to the west of Fan Glacier (ca. 10 km northwest of Qaanaaq town), in Kangerluarsuk/Bowdoin Fjord, at Nuulliit at the entrance to Iterlassuaq/Granville Fjord and around Kangaarsussuaq/Kap Parry. Regarding the latter, the tracks of a female with a little cub were observed in February (the year is unknown) and the hunter expressed surprise over the fact that the family had left the den so early (female polar bears usually leave their dens with their new cubs in March or April, *e.g.* Born *et al.* 1997).

In the southernmost area (Qimusseriarsuaq/Melville Bay), small tracks have been observed close to Innaanganeq/Kap York, Qeqertapaluk/George Island (2006) and Salleq/Bushnan Island (1997).

A 63-year-old hunter from Savissivik remarked that small bear cubs are not seen "around here" (around Savissivik – author's note). However, he and the other hunters had noticed that the small bears have begun to wander down from the north (from outside of Innaanganeq/Kap York) because the "ice does not form properly out there anymore". In March, when the cubs leave the den, one can find tracks leading south, which follow the edge of the land fast ice. According to this hunter, tracks made by very small cubs are never observed in Qimusseriarsuaq/Melville Bay. His assessment of the situation was:

> that before Qimusseriarsuaq was protected in 1974 [sic], we went bear hunting there and caught bear cubs and other young animals. When I was a child, and before the protection was introduced in 1974, we were able to drive dogsleds freely there. For many years

> we drove to Kullorsuaq, and before that area was protected from hunters we saw a lot of tracks made by young bears, but since the area has been protected we no longer see any tracks from bear cubs. This is because the male bears are now left in peace so they can hunt the cubs and eat them. I am quite sure of it. Large bears eat bear cubs [........]. In the days when we used to go bear hunting out on the great sea, we observed a number of times how large males followed the tracks of bears with cubs in March in order to eat the cubs. That has always been the way of things.

The informant confused the protection of female bears with cubs under 1 year old, which was introduced in 1975, with the creation of the Melville Bay Nature Reserve in 1980 (see the chapter "Hunting regulations", p. 23).

A 44-year-old hunter from Qaanaaq town reported having observed the tracks of two small cubs with their mother in the vicinity of Sermersuaq/Humboldt Glacier in April during the 1990s. Their mother had made a ball for them out of a piece of salt water ice, which she had taken from the area between an iceberg and the sea ice. The ball was completely spherical and approximately twice the size of a soccer ball. The hunter described it as follows: "If mathematicians had measured it, they would be amazed at how perfectly round it was." The bear cubs played with the ball (and had slid down the iceberg for a long time). According to the hunter, the ball was probably made of sea ice in order to make it more solid than if it had been fashioned out of freshwater ice from the iceberg which would have been more brittle and would have broken more easily. It was smooth and rounded so that it would not crack and was, in the hunter's opinion, an implement devised and fashioned by a polar bear (see also the section "Have you seen polar bears hunting?", p. 84).

14. Where do the bears spend the summer?

This question was put to 24 hunters, of whom 22 (92%) expressed an opinion on the subject and 2 (8%) had no particular opinion (Table 11).

There were varying opinions as to where the bears are in the summer months (Table 12). A number of the hunters seemed to be unsure of the bears' whereabouts in the summer and several of the responses seemed to be suggestions as to where they might be, rather than actual observations. On the whole, the responses to this question indicated that it is rare to see polar bears in the summer.

Over half of the answers (ca. 59%) indicated that the polar bears are found on the drift ice (or other ice-covered areas) or on land during the summer. Some of the interviewees indicated that the bears prefer glacial areas (Table 12), while one of them was of the opinion that the bears are able to reside anywhere during the summer.

Certain summer areas were specifically identified: Nuussuaq/Washington Land, Nunarsuaq/Inglefield Land, the drift ice in Ikeq/Smith Sound, the western part of Kiatak/Northumberland Island and the drift ice in that area including the mouth of Iluleerloq/Murchison Sound, Qaanaap Kangerlua/Inglefield Bredning around Quinnisut/Hubbard Glacier (observed here recently), the glaciers behind Savissivik and in Qimussiarsuaq/Melville Bay on the glaciers (Mohn, Gade; Döcher Smith and Nordenskiöld Glacier) as well as on the glaciers by Duneira Bay; Fig 13.

Some of the hunters elaborated on their explanations regarding the bears' whereabouts in the summer. Two of them were of the opinion that the bears stay on land primarily when there is little sea ice and another said that the bears hide in the hills because of the noise from skiffs. A hunter from Savissivik explained that the polar

	Total	Siora-paluk	Qaanaaq-Qeqertarsuaq	Savissivik
Opinion on this	22	4	10	8
No opinion	2	1	1	-
N_{total} (hunters asked)	24	5	11	8

Table 11. Distribution of the responses to question 14, from the Qaanaaq municipality: "Where do the bears spend the summer?"

	Total	Siora-paluk	Qaanaaq-Qeqertarsuaq	Savissivik
Everywhere	1	-	-	1
In the drift ice	4	1	1	2
On land	5	2	2	1
Drift ice and on land	4	-	4	-
In the fiords	1	-	-	1
On the glaciers	2	-	-	2
Varies from year to year	1	-	-	1
Kiatak/Northumberland Island	2	-	2	-
Qaanaap Kangerlua/Inglefield Bredning	1	-	1	-
Eastern Ikeq/Kane Basin	1	1	-	-
N_{total} (responses)	22	4	10	8

Table 12. Types of responses to question 14 in the Qaanaaq municipality: "Where do the bears spend the summer?"

FIG. 13.
The polar bears' summering area, as indicated on a map by six hunters during the interviews in the Qaanaaq municipality in February 2006.

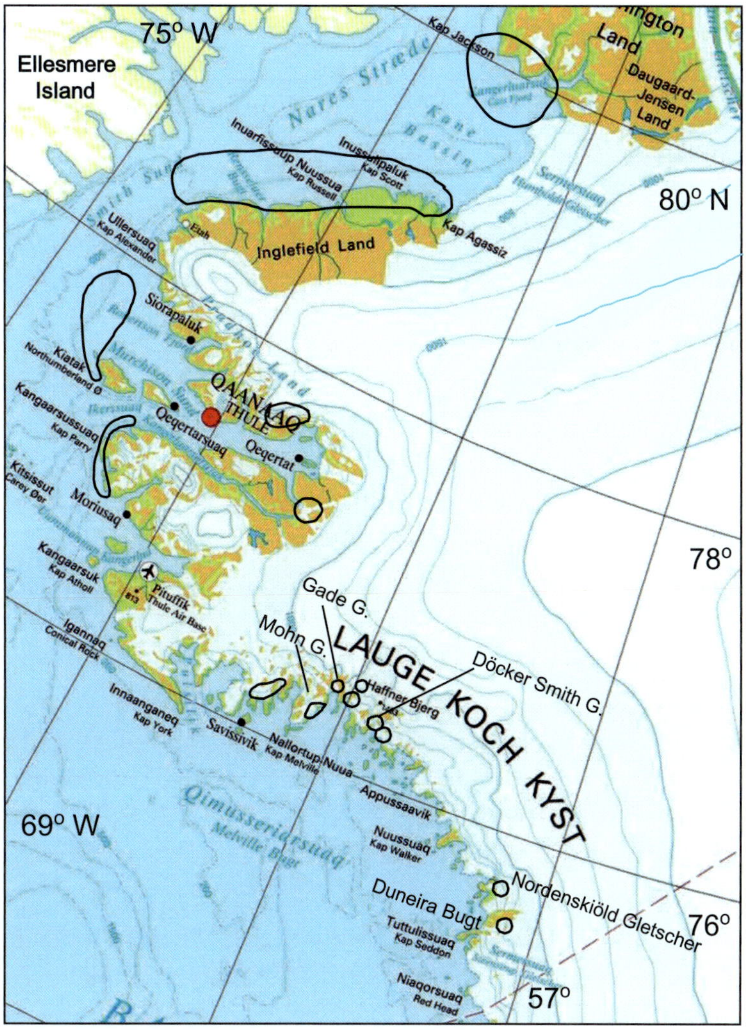

bears tend to be found on or around the glaciers in Qimusseriarsuaq / Melville Bay, as long as there is not too much noise from the skiffs used when hunting narwhals in the area. According to some of the hunters, the polar bears roam around hungry in the summer and may therefore come closer to land, as they are attracted by meat depots.

15. Have you seen dens with females with small cubs (maternity dens)?

Only 3 (12%) of the 25 respondents to this question had seen a maternity den (Table 13). In all three cases, the dens were situated high up on steep mountain sides and had only been observed from a distance. In the spring of 2002, a hunter observed a den close to Inuarfiusapaluk in Nunarsuaq / Inglefield Land. There was evidence of

	Total	Siora-paluk	Qaanaaq-Qeqertarsuaq	Savissivik
Yes	3	2	-	1
No	22	4	11	7
No opinion	-	-	-	-
N_{total} (hunters asked)	25	6	11	8

Table 13. Distribution of the responses to question 15, from the Qaanaaq municipality: "Have you seen dens with females with small cubs (maternity dens)?"

activity around the den, which appeared to be abandoned. Another hunter reported having seen an abandoned den close to Sermitajaaq or Ingeersarfik on Umimaat Nunaat/Ellesmere Island when he was a child (*i.e.* in the 1970s). The tracks indicated that the cub had slid from the den down a snow drift. A third informant had seen a maternity den high up in the hills around Kong Oscars Glacier close to Nuussuaq/Kap Walker in Qimusseriarsuaq/Melville Bay, in around the mid 1950s (Fig. 12).

Two hunters reported on the observations made by others of maternity dens. In both cases the reports were of observations made a long time ago (before 1960) near to Pitoraarfik/Kap Chalon in Murchison Sound and at Qisuup Nunaa (the Benton Bay area east of Kap Clay) on Nuussuaq/Washington Land.

16. Have you seen temporary dens?

Overall, the interviewees had only seen a small number of temporary dens. Of the 24 respondents to this question, 5 (21%) answered "yes" while the remaining 19 (79%) said "no" (Table 14). In one case, however, three of the hunters reported having seen the same den in 2005 (see below).

An adult male polar bear had made a temporary den on the southwestern side of Kiatak/Northumberland Island in 1964 (or 1965); Fig. 12. It had made use of

	Total	Siora-paluk	Qaanaaq-Qeqertarsuaq	Savissivik
Yes	5	-	2	3
No	19	5	9	5
No opinion	-	-	-	-
N_{total} (hunters asked)	24	5	11	8

Table 14. Distribution of the responses to question 16, from the Qaanaaq municipality: "Have you seen temporary dens?"

Tracks of a female polar bear with a yearling near Nuussuaq/Cape Walker in Qimusseriarsuaq/Melville Bay, April 2010. The central part of Melville Bay is a nature reserve to protect denning polar bears. Photo: K. L. Laidre

a total of 5 dens which it had built further and further up the hillside, probably because it had heard the hunters who had come to set their fox traps. The final den, in which the bear was shot just after New Year, was ca. 2 metres in diameter and ca. 1.5 metres high with a long, slightly curved entrance tunnel. An abandoned den was observed by a hunter in the summer of 1949 in a snow drift high up in the hills on Salleq/Bushnan Island near to Savissivik. Three of the hunters spoke of a den which they had observed in August 2005 close to Döcker Smith Glacier near to Fisher Islands in Qimusseriarsuaq/Melville Bay. The den had recently been inhabited by a large male bear that had first dug two other holes in the snow before crawling into the third. The male bear was well fed and had also left excrement around the den. It was shot in the den because the hunters needed meat.

17. Have there been changes to the occurrence of dens over the years?

This question was only put to one of the interviewees as it became apparent during the course of the interviews that only a very small number of dens had actually been observed. A 48-year-old hunter, who had moved to Savissivik from Upernavik some years ago, made the somewhat vague comment that "previously, before we came to live here, the hunters used to see dens", which indicates that there may have been a decrease in the number of dens.

Climate change

18. Have you observed changes to ice conditions?

All of the 25 respondents answered "yes" to this question.

Twenty (80%) of them stated that the sea ice forms later and breaks up earlier. Apart from these statements there were various combinations of responses indicating that the amount of sea ice has decreased and that it is now thinner and less stable with more and larger holes (Table 15).

A hunter from Siorapaluk remarked that he had observed that in recent years the sea ice has not formed until February while another informant said that the ice had begun to form later and break up earlier in the spring since 2000. A 44-year-old informant from the same settlement mentioned that the ice had started to break up

A young female bear at Pitoraarfik/Cape Chalon in the Qaanaaq municipality. Photo: E.W. Born

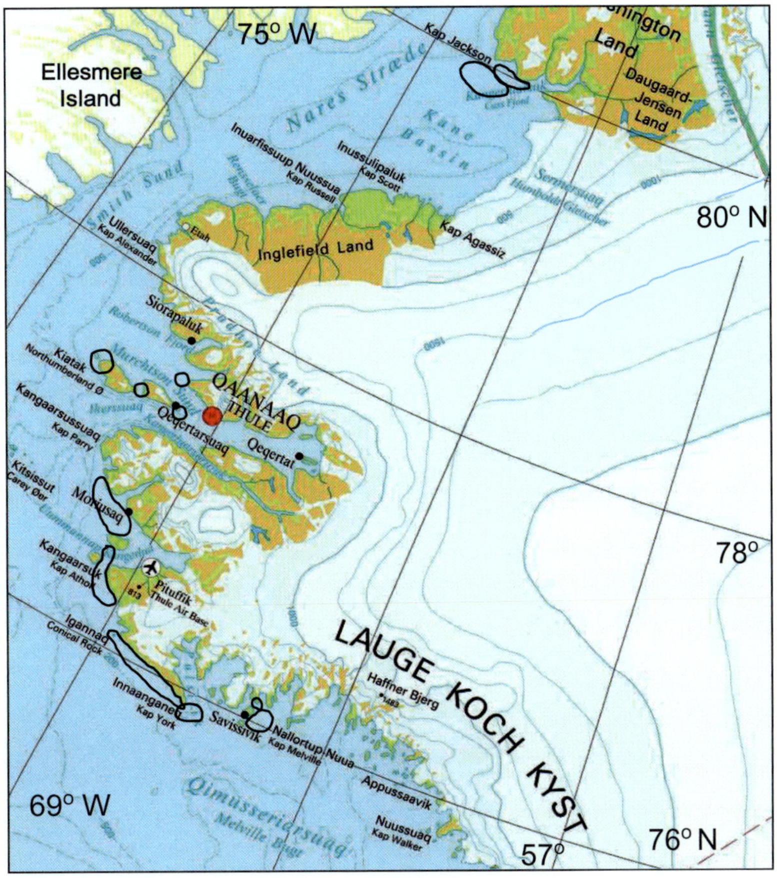

FIG. 14.
Areas with strong currents and holes in the ice as drawn on a map by nine hunters during the interviews in the Qaanaaq municipality in February 2006.

earlier in recent years, and that as a result people tended to use boats more regularly than in the past (they used to travel by dogsled more frequently).

	Number of responses	Siora-paluk	Qaanaaq-Qeqertarsuaq	Savissivik
Forms later and breaks up earlier	20	4	11	5
Thinner and more unsafe	8	2	3	3
Holes in the ice and more open water	6	-	3	3
Changes to the ice edges	4	1	2	1
Other	2	2	-	-
N_{total} (responses)	40	9	19	12

Table 15. Types of responses to question 18, from the Qaanaaq municipality: "Have you observed changes to ice conditions?"

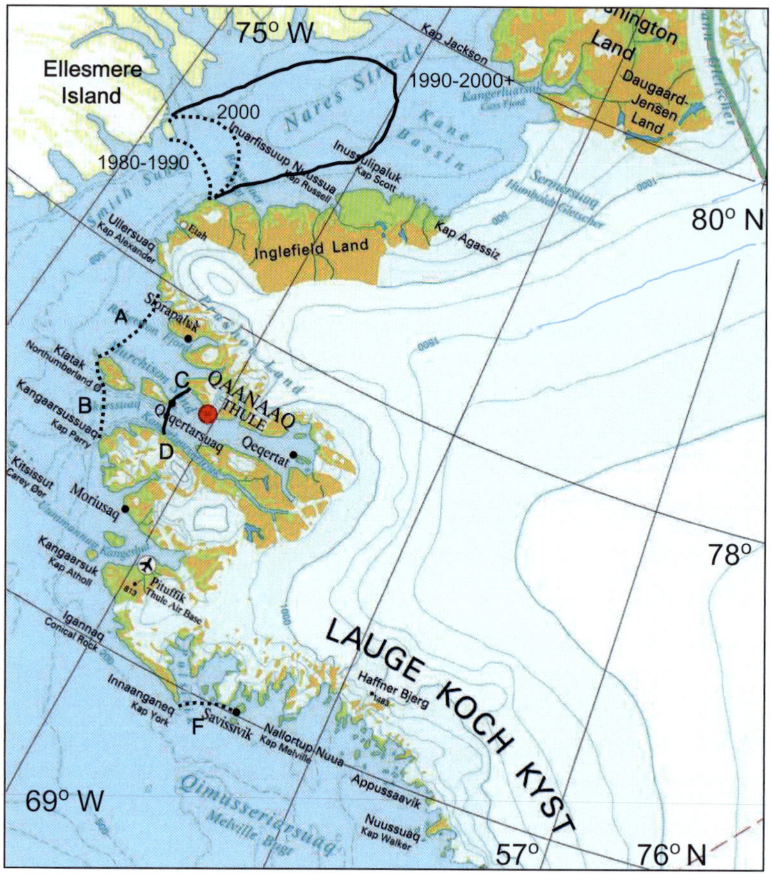

FIG. 15. Former (dotted line) and current (solid line) ice edges, as drawn on a map by eight hunters during the interviews in the Qaanaaq municipality in February 2006. A and B = the position of the ice edge in mid-winter until 1990s; C and D = the position of the ice edge in midwinter (February) since the end of the 1990s; E = former ice edge, where there has been thin and unsafe ice and currents in recent years.

According to the same informant, any hunter who is not going by dogsled via Ikineq at the head of Siorapaluup Kangerlua/Robertsen Fjord would have to use a boat in order to go hunting in the north (*i.e.* in Ikeq and Ikersuaq). He also spoke of how, in his experience and in contrast to what was previously the norm, since 1994 (1995?) large amounts of the heavy pack ice (from Ikersuaq/Kane Basin?) drift in to Iluleerloq/Murchison Sound in the summer. This phenomenon sometimes makes it difficult to travel by boat to Qaanaaq town. A hunter from Siorapaluk said that the ice had started to get "bad" during the 1990s and into the 2000s, while another informant from the same settlement told the interviewers that there is now less ice in the northern part of Ikeq/Smith Sound. According to him, it had become quite difficult to find ice floes which are sturdy enough to flense walruses on in the area.

The hunters from Qaanaaq-Qeqertarsuaq also said that the ice has begun to form later. In the past it was safe to drive a dog team between Qaanaaq and Qeqertarsuaq/Herbert Island even before the polar night set in (*i.e.* probably meaning that one could drive in October; author's note). In recent years this has not been possible until January.

In addition to this, in 2004 and 2005 it was not possible to drive over the ice to Sioratooq (situated about 20 km to the west of Qaanaaq) until the end of December. However, it was mentioned that if the icebergs run aground on the reef just outside of the town of Qaanaaq fast ice still forms between this reef and the coast. One informant remarked that in the past there had still been daylight when the new ice formed and when the seal hunting on the thin ice took place. At that time it was also possible to travel by dogsled along the coast around Kangaarsussuaq/Kap Parry in December.

A 39-year-old interviewee who had lived in Savissivik as a child (*i.e.* at some point in the 1970s) recalled how the ice used to form in October during his childhood, but said that nowadays it does not form until January. Another informant from the same settlement said that the ice actually forms in December but is not safe to travel over until January, while a third informant stated that the sea ice tends to form as late as February. According to one informant this change in timing of the formation of new ice occurred in the Savissivik area during the 1980s and 1990s. The same man added that the sea ice had not formed "properly" since 2000. The ice is now said to break up at some time in June, whereas in the past this tended to occur as late as August.

Several informants mentioned that the sea ice has also become less safe for travelling and hunting because it is thinner and therefore breaks up more frequently during stormy weather (Table 15). This may happen several times throughout the winter and, following such stormy weather, the ice may take a long time (sometimes weeks) to form again because of waves and swells. Hunters from each of the areas in this study cited this as the reason why travelling over the ice has become increasingly dangerous. A 44-year-old hunter from Siorapaluk said that the ice is now ca. 0.5 m thick whereas it used to be 1.5–2 m thick in the winter. This informant also described how the southerly wind can easily break up the ice, a phenomenon that he experienced while hunting walruses in the area between Pitoraarfik and Kiatak. A 64-year-old from the same settlement described the ice as follows:

> It doesn't get thick. It used to get very thick, as thick as a man's height. Now it is very thin … less than a metre.

According to this man, this phenomenon is due to an increase in the strength of the current. A 40-year-old hunter who had moved from the settlement of Kullorsuaq in the Upernavik municipality to Savissivik said that the ice to the east of Kullorsuaq has become markedly thinner in recent years. Whereas it used to be more than a metre thick, it now only measures about 25 cm. A 63-year-old hunter from Savissivik said that the ice in the area had been very thick and safe in the past and that cracks in the ice used to refreeze in the winter, creating areas of thinner ice where ringed seals could make breathing holes. Nowadays these cracks widen greatly in a short space of time and when they freeze over, the ice is so unsafe it is impossible to drive over it to reach the safe ice on the other side. He also mentioned that they usually travel by dogsled to the bird areas in order to catch little auks (*Alle alle*) and gather eggs (15 June). This journey was impossible in 2005 because the cracks in the ice were too wide.

The instability of the ice was also ascribed to the fact that there is now more open water, a greater number of places where there is a strong current and more holes in the ice. The size of the holes has also increased (Table 15). These conditions make travelling by dogsled more difficult and more dangerous. The holes in the ice remain open no matter how cold it gets and one informant described how the ice around the holes is less than half a metre thick because of the current. The holes in the ice by Kiatak/Northumberland Island and Qeqertarsuaq/Herbert Island have increased in size and have a stronger current, and there is now open water and unsafe ice in the Moriusaq area (*e.g.* off the coast at Nuulliit and Qooqqut) and along the coast of Sanerarsuaq between Kangaarsuk/Kap Atholl and Innaanganeq/Kap York (Fig. 14). There is now also open water and a strong current around Innaanganeq /Kap York. Between this cape and Savissivik, where there once was fast ice with a well defined edge, there is now thin ice with holes in a number of areas. It was mentioned that holes have also formed in the ice near to Assorput, on the south side of Savissivik/Meteorit Island, upon which the settlement Savissik is situated. Furthermore, holes have appeared in the ice near to the island's westernmost point at Nuussuaq, as well near Kangilipaliit on the eastern side of the island. The ice in Ikera/Meteorit Bugt to the east of the island has now also become thin and unsafe (Fig. 14). A hunter told how it has been impossible since the late 1990s to drive a dog team out in a westerly direction "offshore" because the sea ice has been so unsafe and because there is now so much open water.

As a result of changes to ice conditions, the location of the edges of the ice has also shifted a great deal. A hunter from Siorapaluk stated how the southernmost edge of the pack ice in Ikersuaq/Kane Basin has been situated further north since the 1990s due to a reduction of ice in Ikersuaq (Fig. 15). According to several of the informants during the 1990s, the edge of the land fast ice, which used to be situated at the mouth of Iluleerloq/Murchison Sound and Ikeq/Hval Sund, has shifted markedly towards the east. In February the ice edge used to lie just off the most westerly point of Kiatak/Northumberland Island, yet in recent years has been located close to the easternmost point of Qeqertarsuaq/Herbert Island (Fig. 15).

19. Have you seen changes to the icebergs?

Of the 25 interviewees who were asked this question, 20 (80%) answered "yes", and the remaining 5 (20%) answered "no" or that there had been no change (Table 16).

Dog sleds drive along the ice foot next to land when there is too much open water or unsafe sea ice. Photo: M. V. Jensen

	Total	Siora-paluk	Qaanaaq-Qeqertarsuaq	Savissivik
Yes	20	4	8	8
No/Same	5	2	3	-
No opinion	-	-	-	-
N_{total} (hunters asked)	25	6	11	8

Table 16. Distribution of the responses to question 19, from the Qaanaaq municipality: "Have you seen changes to the icebergs?"

	Number of responses	Siora-paluk	Qaanaaq-Qeqertarsuaq	Savissivik
Decreased in number	16	3	5	8
Increased in number	2	1	1	-
Other	3	-	3	-
N_{total} (responses)	21	4	9	8

Table 17. Types of responses to question 19 in the Qaanaaq municipality: "Have you seen changes to the icebergs?"

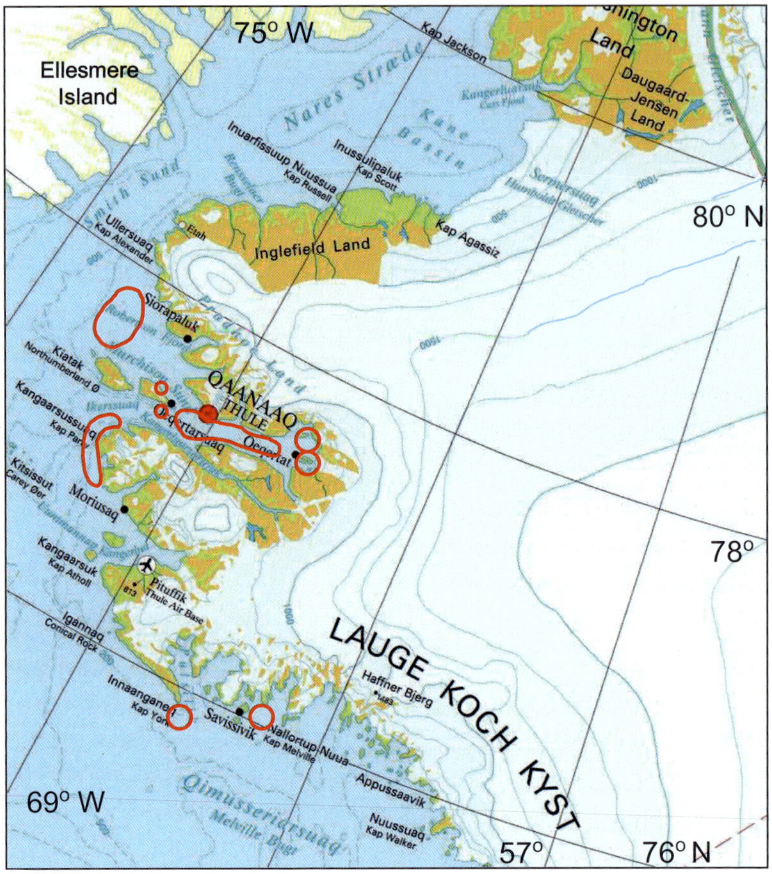

FIG. 16.
Areas where there are fewer icebergs, as indicated on a map by 20 hunters during the interviews in the Qaanaaq municipality in February 2006.

The majority of the responses (ca. 76%) indicated that there are now fewer icebergs than before (Table 17). Two responses indicated however that the number of icebergs had increased, but both cases referred to specific areas.

Two hunters from Siorapaluk reported that there has been a decrease in the number of icebergs that run aground on the bank just offshore of Neqi – an area called Siattorissat (Fig. 16). According to a 44-year-old highly experienced hunter from that settlement, there are still many icebergs in Ikersuaq / Kane Basin whereas in the area in front of Sermersuaq / Humboldt Glacier in Ikersuaq / Kane Basin they are now more scattered. A 47-year-old hunter from the same settlement recalled how when he had begun to hunt polar bears (in 1989; author's note) there were only a few icebergs in Ikeq / Smith Sound and in Ikersuaq / Kane Basin north to Nuussuaq / Washington Land, but since 1998–1999 a greater number are found in the eastern parts of Ikersuaq (*i.e.* in front of Sermersuaq).

The hunters from Qaanaaq town and Qeqertarsuaq also indicated that there has been an overall decrease in the number of icebergs. According to one of them, marked changes have taken place around the two locations named Upernavissuit, located on the north and south side of Qeqertarsuaq/Herbert Island respectively, and that icebergs seldom come to these areas anymore. The same man also explained that while icebergs still drift over in a westerly direction from Qaanaap Kangerlua/Inglefield Bredning, there are no longer any "jagged" icebergs (large icebergs with a jagged top, which break off from glaciers; author's note). This statement contrasted however with that of another informant, who maintained that a large number of jagged icebergs flow out from the eastern part of Qaanaap Kangerlua. However, generally there are now fewer icebergs around the glaciers at the base of Qaanaap Kangerlua (Fig. 16).

One hunter remarked that due to the late freeze-up, the icebergs that people so hope to see run aground in front of Qaanaaq town drift away, much to everyone's dismay (stranded icebergs serve to provide fresh water, keep the fast ice in place, and set seal nets near them; author's note). According to the information we gathered, it appears that hardly any icebergs run aground on the shores at Kangaarsussuaq/Kap Parry anymore. These icebergs used to keep the land fast ice in place so it did not break up and disappear early in the season.

Six out of the eight informants from Savissivik mentioned how there are now far fewer icebergs in Ikera/Meteorit Bugt to the east of the settlement (Fig. 16). Two of these respondents mentioned that in the 1970s and 1980s, there were so many icebergs that it was impossible to pass through them on a dogsled. One informant remarked that fewer icebergs results in more powerful swells in the summer because the icebergs are no longer around to dampen the motion of the sea. A hunter from Qaanaaq town with hunting experience in the Savissivik area was however of the opinion that overall the number of icebergs in the Savissivik area had remained unchanged, but that changes in the current patterns have meant that the icebergs no longer gather in Ikera/Meteorit Bugt. One informant also stated that there are now fewer icebergs in the vicinity of Innaanganeq, while another said that this also applies to *"Qorfiit"* (*"The Pots"*, *i.e.* an offshore area of shallow water, which used to be an important bear hunting area before the effects of climate change set in; author's note; Fig. 2).

20. Have you seen changes to the glaciers?

Of the 24 informants who were asked this question, 22 (92%) were answered "yes", while 2 (8%) expressed no specific opinion on the subject (Table 18).

All of the 22 informants who answered this question in the affirmative spoke of how the glaciers have receded and how sections of the Inland Ice Cap have shrunk or decreased in size. However, five of the informants were also able to pinpoint

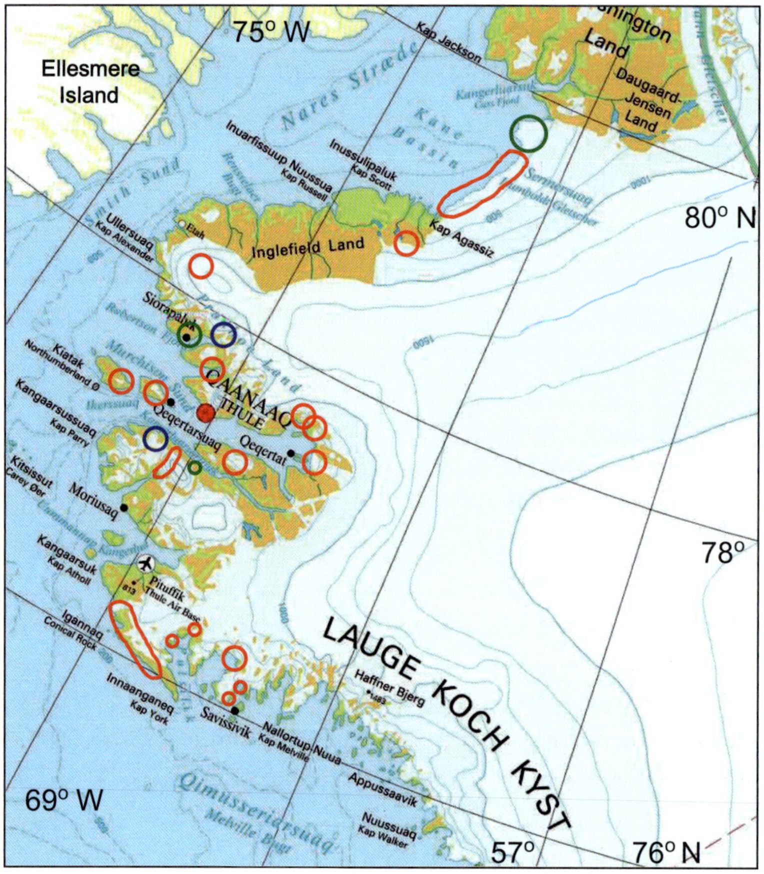

FIG. 17. Recent changes to glaciers and parts of the ice cap, as drawn on a map by 20 hunters during the interviews in the Qaanaaq municipality in February 2006. Red = glaciers that have receded; blue = glaciers that have receded but that have increased in productivity; green = glaciers that have increased in productivity.

certain glaciers whose productivity appears to have increased in recent years (Table 19).

A 64-year-old hunter from Siorapaluk told of how the glacier in the vicinity of the settlement has shifted 100 metres closer to the coast and increased in productivity in the last 30 to 40 years. In contrast to this, the glacier in the inner part of Siorapaluup Kangerlua / Robertson Fjord has shrunk but is still nonetheless productive. According to another informant, the glaciers at Neqi and Ullersuaq / Kap Alexander have receded and the section of the Inland Ice Cap that is located in the vicinity of Qaqaitsut / Kap Agassiz in the eastern part of Nunarsuaq / Inglefield Land has also decreased in size (Fig.17). One of the hunters was of the opinion that in overall terms, the glacier fronts have receded 500–1000 m from the coast since 2000.

Several of the hunters from Qaanaaq-Qeqertarsuaq reported that the glacier ("that has serrated or jagged sections on its edges"), located east of Qeqertaarsuusarsuaq / Josephine Peary Island in the inner part of Qaanaap Kangerlua / Inglefield Bredning, has shrunk to such an extent that it is now possible to circumnavigate Josephine

	Total	Siora-paluk	Qaanaaq-Qeqertarsuaq	Savissivik
Yes	22	5	10	7
No / Same	-	-	-	-
No opinion	2	-	1	1
N_{total} (hunters asked)	24	5	11	8

Table 18. Distribution of the responses to question 20, from the Qaanaaq municipality: "Have you seen any changes to the glaciers?"

	Number of responses	Siora-paluk	Qaanaaq-Qeqertarsuaq	Savissivik
Retreated	22	5	10	7
More productive	5	1	2	2
N_{total} (responses)	27	6	12	9

Table 19. Types of responses to question 20 in the Qaanaaq municipality: "Have you seen any changes to the glaciers?"

Peary Island (*i.e.*, sail to the east of the island between the island and the Inland Ice Cap). This was not possible a few years ago. In addition to this, the informants stated that the calving of icebergs in that area now causes large swells that have begun to disrupt narwhal hunting and on occasion destroy kayaks on the shore. Another conspicuous development, which was named by several different informants, is that Itillersuaq/Politikens Bræ (Politiken's Glacier) is no longer passable or is barely passable. Melt water run off has created large fissures in the edge of this glacier close to Ikeq/Hval Sund. This glacier used to be the main thoroughfare between the southern and northern parts of the district. According to two other informants, the glacier in the inner part of Iterlassuaq/MacCormick Fjord and a glacier on the north side of Qeqertarsuaq/Herbert Island have also receded. The same has happened to several glaciers along Sanerarsuaq between Kangaarsuk/Kap Atholl and Innaanganeq/Kap York. Furthermore, one informant spoke of how the ice cap on the land between Qaanaap Kangerlua/Inglefield Bredning and Kangerluarsorujuk/Olrik Fjord has decreased in size, and that as a result large vegetation-free sections of land have been exposed.

A number of smaller glaciers are however highly productive in spite of the fact that they have decreased in size. Some examples of this phenomenon are the Melville, Sharp and Hart glaciers north of Qeqertaarsuusarsuaq/Josephine Peary Island, Misuumasoq and Mitchell's Bluff, as well as Kiggaviarupaluk west of Itillersuaq/Politikens Bræ.

Seven hunters from Savissivik spoke of receding glaciers, as well as some glaciers which are still active and have increased in productivity. According to one informant, the glaciers in Ikera/Meteorit Bugt are no longer productive, and the glaciers in Illaarsuk/Sidebriks Fjord have shrunk, whereas the glaciers in Puisilluusarsuaq/De Dødes Fjord and Gade Glacier have remained unchanged and are still productive. Sermipaluk Glacier behind Akuliaruseq has dramatically decreased in size (since 1975?) and the glacier at Eqalutik has also been reduced. However, according to the same informant, the glacier at Qeqertaq/Salve Island has started to produce icebergs.

Changes to the glaciers have (among other things) meant that the sled route to Pituffik via Puisilik is no longer passable due to leads in the edge of the glacier and because the Inland Ice Cap has decreased in size in that area.

21. Have you seen any changes to the snow?

Nineteen (83%) of 23 interviewees answered "yes" to this question and 4 (17%) answered "no", or that "there had been no change" (Table 20).

The majority of the informants had observed a reduction in the amount of snow in recent years and some responses indicated that there had also been a change with respect to when the snow falls (Table 21).

	Total	Siora-paluk	Qaanaaq-Qeqertarsuaq	Savissivik
Yes	19	5	8	6
No/Same	4	-	3	1
No opinion	-	-	-	-
N_{total} (hunters asked)	23	5	11	7

Table 20. Distribution of the responses to question 21, from the Qaanaaq municipality: "Have you seen any changes to the snow?"

	Number of responses	Siora-paluk	Qaanaaq-Qeqertarsuaq	Savissivik
Less falls	16	4	8	4
Change in the time of year	6	2	-	4
N_{total} (responses)	22	6	8	8

Table 21. Types of responses to question 21 in the Qaanaaq municipality: "Have you seen any changes to the snow?"

All of the interviewees from Siorapaluk were of the opinion that there had been a decrease in snowfall in recent years. One informant said that in a "bad" winter, there is heavy snowfall in the spring and the snow comes later in the autumn. According to this informant, less snow currently falls in May. Another informant supported this statement, saying that there is now more snowfall in March and April as opposed to in May. This means, among other things, that it is difficult to drive a dog team over the ice because the snow conceals holes in the ice and other dangerous areas. He added that heavy snowfall early in the year also complicates the walrus hunt on the newly formed ice because the snow squeaks when the hunters walk softly on the thin ice towards walruses in order to harpoon them.

Eight of the 11 informants from Qaanaaq said that it snows less now than in the past. A 49-year-old hunter added that there was more snow during his childhood in the 1960s. As in Siorapaluk, the informants indicated that the snow now falls later in the year and not until the spring. One interviewee recalled that he had previously been able to build a shelter from the wind out of snow in the winter when he went out on the ice to catch Greenland halibut (*Reinhardtius hippoglossoides*) and said that this is no longer possible. Two hunters emphasised that the lack of snow may cause problems for the sled dog's water supply. One of them told the interviewer this means that it is necessary for him to take a sack of snow in order to provide his dogs with fresh water when he goes hunting in the autumn.

A 66-year-old hunter from Savissivik mentioned there had been a great deal more snow when he was a child. In the Savissivik area, the spring snowfall has come later in recent years and there has been a reduction in the overall amount of snow. The statements of two hunters indicate that the snowfall now fluctuates more and is less predictable. One informant said that there had been a large amount of snow a couple of years ago, while another mentioned that there had been a lot of snow in the preceding year and the year before that (*i.e.* 2004 and 2005), but that currently, there was on the whole less snow. A 48-year-old hunter recalled that when he was a child the snow used to fall in the period of calm following a storm, but that nowadays snowfall and wind storms can occur simultaneously (see also the answers to question 23, p. 68).

However, four hunters had the opinion that the snowfall had not changed (3 from Qaanaaq, 1 from Savissivik). Two of these hunters stated that the amount of snow varies from year to year, while one of them (Savissivik) said there used to be more snow in the spring. A 54-year-old hunter from Qaanaaq town stated that there had never been much snowfall in the Qaanaaq area, whereas a 69-year-old from the same town said that there had always been frequent snowfalls in the area (this hunter had, however, lived in many different places in the municipality; author's note).

The settlement of Savissivik in February 2006. Photo: T. Qaavigaq

22. Have you seen changes to current patterns?

Of the 23 respondents, 17 (74%) answered "yes" to the question, 5 (22%) answered "unchanged" while 1 (4%) had no opinion on the subject (Table 22).

The majority of the responses made it clear that the currents have increased in strength, but some also suggested that the currents have changed direction (Table 23). An informant from Siorapaluk noted that the current in Iluleerloq / Murchison Sound-Qaanaaq area had grown stronger and that as a result the ice has become thinner with holes. Another interviewee (aged 46) was of the opinion that the current patterns started to change around 2003. Moreover, he had observed that the tidal variations were more extreme when there was a new or a full moon. This phenomenon (higher water at high tide) had also been observed by a hunter from Qaanaaq town. Moreover, the 46-year-old hunter from Siorapaluk said that in the summer of 2005, during a hunting trip to Qeqertat / Howard Islands in the inner part of Qaanaap Kangerlua / Inglefield Bredning, he had seen a huge wave which had freed the stranded icebergs from the ocean floor. The giant wave moved back and forth about 10 times and destroyed hunting implements on the shore. The wave appeared to have emanated from the west and did not appear to be due to a calving glacier or a capsized iceberg. The hunter suggested that the wave had been the result of a shift in the ocean floor ("tsunami").

Several of the hunters from Qaanaaq town mentioned that the current had increased in strength and that the holes in the ice had expanded, and one of them said that water

	Total	Siora-paluk	Qaanaaq-Qeqertarsuaq	Savissivik
Yes	17	3	7	7
No / Same	5	1	3	1
No opinion	1	-	1	-
N_{total} (hunters asked)	23	4	11	8

Table 22. Distribution of the responses to question 22, from the Qaanaaq municipality: "Have you seen any changes to current patterns?"

	Number of responses	Siora-paluk	Qaanaaq-Qeqertarsuaq	Savissivik
Have become stronger	12	2	6	4
Have changed direction	4	-	1	3
Other	2	1	-	1
N_{total} (responses)	18	3	7	8

Table 23. Types of responses to question 22 in the Qaanaaq municipality: "Have you seen any changes to current patterns?"

sometimes flows over the ice when there is a strong current and holes in the ice pushing it down and causing it to break up, even in areas where the ice is not particularly thin. The areas of strong current at Kiatak/Northumberland Island and Qeqertarsuaq/Herbert Island are now stronger and larger than they were previously (Fig.14). There are now areas of strong current and unsafe ice in the Moriusaq area, especially in the Nuulliit area to the west of the mouth of Iterlassuaq/Granville Fjord. One of the hunters stated that the location of the areas of strong current have changed and that their occurrence is more unpredictable than it was in the past, when their location was always well-known.

The hunters from Savissivik had also observed that the current had increased in strength and one of them was of the opinion that the change occurred at the beginning of the 1990s. In order to describe the strength of the current, the same man explained that the stones, used to hold the seal nets in a vertical position in the water, have now been seen "floating" in a horizontal position because of the current. The ice between Savissivik and Serfarmiut in the Innaanganeq/Kap York area is now unsafe and it is dangerous to drive a dog team over because of the current. According to the informants, the current creates areas of thin ice and holes in the ice in unexpected locations, such as in between icebergs. This occurs in Ikera/ Meteorit Bugt to the east of Savissivik, in particular. One informant spoke of how the increased strength of the current and the resulting undercut ice means that it is dangerous to drive a dog team over the ice in the spring (Fig.14).

23. Have you seen any changes to the weather?

Of the 23 informants to whom this question was posed, 20 (87%) answered "yes", 2 (9%) answered "no", while 1 (4%) had no particular opinion on the matter (Table 24).

Hunters from all of the areas in the municipality had experienced that the weather had in general become warmer, stormier, and more unstable (Table 25). Six informants expressed the opinion that the temperature had increased, even in winter, and that there has been more thawing at unusual times of the year. A hunter from Qaanaaq thought that the changes had begun in the late 1990s and added that the temperature seldom dropped to ca. -40°C. A hunter from Savissivik said that there has been a marked increase in the temperature in the spring, but in his opinion, the winter temperature is still as low as in the past. Several responses to this question indicated that the winter of 2005/2006 was relatively cold and stable, also similar to the past.

Approximately 38% of the responses indicated that the weather had become stormier (Table 25). According to one hunter from Siorapaluk, weather systems have become more extreme, and another hunter from the same settlement said that the weather had been especially bad in the period 1998–2004. A 43-year-old resident of Siorapaluk said:

> The wind blew constantly in the 1980s. In 2004, there was a perpetual southerly wind, there was no calm weather at all that whole summer. In the end, we had nothing at all and in 2005 we were short of food. Some years are like that. As our ancestors said it has been this way every 100 years. They went hungry then, for example, just like what we experienced in 2004.

	Total	Siora-paluk	Qaanaaq-Qeqertarsuaq	Savissivik
Yes	20	5	9	6
No/Same	2	-	2	-
No opinion	1	-	-	1
N_{total} (hunters asked)	23	5	11	7

Table 24. Distribution of the responses to question 23, from the Qaanaaq municipality: "Have you seen any changes to the weather?"

	Number of responses	Siora-paluk	Qaanaaq-Qeqertarsuaq	Savissivik
Warmer	6	1	3	2
More wind	9	2	5	2
Unstable/rainy	7	1	3	3
Other	2	-	2	-
N_{total} (responses)	24	4	13	7

Table 25. Types of responses to question 23 in the Qaanaaq municipality: "Have you seen any changes to the weather?"

A hunter from Qeqertarsuaq stated that the strong wind can rise very quickly and an informant from Qaanaaq said that there is noticeably more wind in the autumn, in particular an easterly wind from Qaanaap Kangerlua/Inglefield Bredning. He also mentioned that both the northerly and southerly winds are now very powerful. Another Qaanaaq resident said that it is very windy in the summer. In the same town, a 64-year-old respondent mentioned how the weather had been calmer in the autumn when he was a child and that in addition to this, the wind has now changed direction. Furthermore, according to this informant, there are now more instances of wind accompanied by snowfall and more days when the sky is cloudless in spite of the presence of a southerly wind. The same man also voiced the opinion that it had become less windy around Qeqertarsuaq/Herbert Island. The combination of snowfall and wind at the same time was also mentioned by an informant from Savissivik. The same man also said that in the autumn wind blows causing high waves at the time of year when the new ice usually forms. Informants also indicated that the wind comes from a different direction than it used to.

Around a third of the answers addressed the unstable weather conditions, mentioning among other things the increase in rainfall and/or that the rain falls at unusual times of year (Table 25). An informant from Siorapaluk told the interviewer that the weather has become more unpredictable and that the northerly wind is less common now than it was in the 1980s and 1990s. Three hunters from Qaanaaq-Qeqertarsuaq mentioned that it rains more now. One of them said that the spring and summer weather is more unstable whereas the "winters are fine". This informant added that he thought that the snow which should have fallen in the winter now falls as rain in summer instead. One of the interviewees from Savissivik said that weather conditions in November and December have become more changeable. Another hunter from the same settlement told the interviewer that there are times when a few cold days are followed by warm weather, while another stated that the weather is sometimes fine for a couple of days after which it is bad for a week or more.

Two hunters from Qaanaaq were of the opinion that the weather had not undergone changes, and one of them expressed this sentiment as follows:

> But clearly, all years can not be the same. Some years it turns out as it ought to, like this year for example (the winter of 2005/2006; author's note).

24. Have you observed any other changes?

Of the 23 respondents, 14 (61%) answered "yes" and 9 answered "no" (Table 26).

The intention with this question was to glean information about any other changes which may be related to climate change. The responses, however, contained information on a variety of subjects (Table 27).

Some of the informants spoke of changes that have affected the walrus hunt. A hunter from Siorapaluk told how the ice around him had broken up when he had been out hunting walruses on the thin ice off the coast by Pitoraarfik with other hunters in 2005. This occurred because the ice is no longer as thick as it used to be and is easily broken up by the southerly wind. In the 1970s and the early 1980s, hunters used to go walrus hunting on the thin ice when there was a southerly breeze because that was when it was least dangerous. However, nowadays it has become

The 332 m high island Usuussarsuaq/Melville's Monument in Qimusseriarsuaq/Melville Bay, April 2010. Photo: K. L. Laidre

	Total	Siora-paluk	Qaanaaq-Qeqertarsuaq	Savissivik
Yes	14	4	4	6
No	9	-	7	2
N_{total} (hunters asked)	23	4	11	8

Table 26. Distribution of the responses to question 24, from the Qaanaaq municipality: "Have you observed any other changes?"

	Number of responses	Siora-paluk	Qaanaaq-Qeqertarsuaq	Savissivik
Walrus catch	4	1	2	1
Migration routes	3	1	-	2
Bald seals	4	2	-	2
Other	6	1	3	2
N_{total} (responses)	17	5	5	7

Table 27. Types of responses to question 24 in the Qaanaaq municipality: "Have you observed any other changes?"

hazardous to hunt on the thin ice when the wind comes from the south because the wind causes the ice to break up. The same man stated that during the 1990s, there had been a large amount of heavy ice floes in the autumn when hunters went out among the drift ice to hunt walruses. Nowadays one sees fewer heavy ice floes in the walrus hunting areas (*i.e.* in the mouth of Iluleerloq/Murchison Sound and in the Kiatak area; author's note) and on occasion finding an ice floe on which to butcher the animals can be problematic. Another informant (aged 64) told the interviewer that he did not know of anyone who had gone walrus hunting on the thin ice in the mouth of Murchison Sound in recent years in spite of the fact the there are a large number of walruses on the mollusc banks in that area. This is due to the changes in ice conditions in the area (most likely because stable new ice with walrus breathing holes in it no longer forms; author's note). In 2005, no new ice formed at all in the area where hunters usually go walrus hunting on thin ice. The same man also indicated it may be that the walruses leave the area quickly (*i.e.* the mouth of Murchison Sound in the Kiatak area) because of the noise from outboard motors. A hunter from Siorapaluk reported that when the hunters go out to walruses in May, the animals disappear off in a northerly direction soon afterwards because they have grown more timid. This did not happen in the 1980s, when it had been possible to catch many walruses in that area in May and June and, on occasion, even in July.

Three informants mentioned changes to the migration routes of belugas (*Delphinapterus leucas*), narwhals, and walruses. A 47-year-old resident of Siorapaluk said that the belugas have chosen a new route and no longer pass by *"Avanersuaq"* (*"The great or far North"*, *i.e.* the Qaanaaq municipality) and that this is apparently due to the noise made by outboard motors. Migration routes (probably in the autumn; author's note) go much further south than they did in the past, according to this informant. A 48-year-old hunter from Savissivik mentioned that the narwhals used to come closer to the settlement during their migration to the south in order to feed on Greenland halibut (his explanation indicates that the narwhals used to follow the edge of the stable land fast ice; author's note). However, the drift ice, which floats around off the coast of Salleq/Bushnan Island, now prevents them from doing this. This heavy drift ice originates in Qimusseriarsuaq/Melville Bay where it is broken up by the waves, and the informant believed that as a result the narwhals follow a more westerly route along the eastern edge of the drift ice area (*i.e.* further out to sea). As a result of this phenomenon, there has apparently been a reduction in the narwhal catch in recent years. A 63-year-old hunter was however of the opinion that the narwhals had avoided the Savissivik area for a number of years because they had been hunted from cutters by hunters from communities to the south who travelled to the ice-edge as far north as Qeqertaq/Salve Island and Nallortoq by Kap Melville. According to the same man, the narwhals have returned in spring and summer thanks to hunting restrictions. Additionally, the same informant mentioned that the number of walruses along the ice-edge around Salleq/Bushnan Island in the spring has decreased. According to him, the walruses have changed their route possibly because the offshore ice has broken up, so they can wander out further.

Bald seals or seals with a strange coat were also mentioned by the interviewees. A 40-year-old hunter from Savissivik mentioned how some very lean young harp seals (*Phoca groenlandica*) had been caught a few years ago at Kullorsuaq (and later in Savissivik) and that they had shed their fur which meant that the skins could not be sold. He added that one no longer hears of such cases. A 64-year-old hunter from Siorapaluk said that since the ice had begun to form later, ringed seals have on occasion been observed moulting as though they come from somewhere else (this probably means that the time of year when they moulted was unusual for seals in the Qaanaaq municipality; author's note.) Their fur was brown in colour and therefore could not be sold. A similar phenomenon was reported by a 48-year-old hunter from Savissivik. He and some other hunters had observed that some of the seals caught in nets have very poor yellowish-brown fur. Also, in the summer of 1997 (or 1998), he caught a ringed seal that was completely hairless ("as though it had been plucked") except around its lips.

On the subject of "other changes", one 44-year-old hunter from Qaanaaq spoke of how surfaces on land are covered with a crust of ice when a longer period of thawing is followed by a decrease in temperature. He thought that this adversely affects herbivores such as ptarmigan (*Lagopus mutus*), Arctic hares (*Lepus arcticus*) and reindeer (*Rangifer tarandus*), which subsequently grow very thin. This has resulted in a reduction in populations of these species in recent years. According to a 46-year-old from Siorapaluk, the number of reindeer in Nunarsuaq/Inglefield Land has declined steeply. There was apparently a migration from Qaqqaitsut/Kap Agassiz in Inglefield Land south to Siorapaluk in 2001 (or 2002), and a large number of reindeer died in 2003 (or 2004) because there were too many animals in the vicinity of Etah/Foulke Fjord. In his opinion, the changes may have been the result of over-grazing due to an increase in the population caused by hunting restrictions in the area since 1987/1988, as well as the increase in competition for food due to the introduced musk ox, *Ovibos moschatus* (introduced in 1986; author's note). This informant also philosophised that the reindeer might also be dying because they accidentally swallow musk ox hair, which causes their stomach to swell and rupture. The same man also said walruses and bearded seals (*Erignathus barbatus*) are sometimes found dead even though they are not wounded in any way. He did not know where these walruses and bearded seals came from but said that they floated round in the water in an advanced state of decay and smelled awful.

A hunter from Savissivik mentioned that ringed seals have started to come to the area in order to haul out on the ice. In his opinion, this was because there was no longer any sea ice on which they could bask further offshore.

A couple of other hunters' responses to this question included the statement that polar bears have started to come closer to the coast or regularly used hunting areas because of the reduction in ice (see question 25, p. 74).

Dogs sleds are used for transporting skiffs to the ice edge where they can be launched during spring. Photo: M. V. Jensen

Changes to the catch and occurrence of polar bears

25. Have the changes affected the polar bear hunt?

This question referred specifically to the extent to which climate-related changes (sea ice, wind, glaciers, weather etc.) have affected polar bear hunting. Of the 22 hunters who were asked this question, 20 (91%) answered "yes" and 2 (9%) answered "no" (Table 28).

	Total	Siora-paluk	Qaanaaq-Qeqertarsuaq	Savissivik
Yes	20	4	8	8
No / same	2	1	1	-
N_{total} (hunters asked)	22	5	9	8

Table 28. Distribution of the responses to question 25, from the Qaanaaq municipality: "Have the changes affected the polar bear hunt?"

About 54% of the responses stated that the bears have, in general, come closer to the coast. Around 31% (n = 8) of the affirmative responses indicated that the changes to the distribution are the result of changes to the ice conditions. Circa 39% of the responses to this question suggested that there have been changes to the routes taken during bear hunts (Table 29).

Regarding the changes to hunting routes, one hunter from Siorapaluk said that hunters used to drive north to Nuussuaq / Washington Land in Ikersuaq / Kane Basin to hunt polar bears, whereas nowadays they only had to go as far as Nunarsuaq/

	Number of responses	Siora-paluk	Qaanaaq-Qeqertarsuaq	Savissivik
Routes changed	10	2	3	5
Closer	6	1	1	4
- due to changes in the ice	8	1	4	3
Changes to methods	1	1	-	-
Other	1	-	1	-
N_{total} (responses)	26	5	9	12

Table 29. Types of responses to question 25 in the Qaanaaq municipality: "Have the changes affected the polar bear hunt?"

Inglefield Land in southern Ikersuaq/Kane Basin because the bears have come closer. Another hunter from Siorapaluk said however that the reduction in ice in Ikeq/Smith Sound and Ikersuaq/Kane Basin has resulted in a decline in hunting activity (*i.e.* using dogsleds) because the ice is no longer safe. The fact that hunters have not travelled north to hunt bears in recent years was confirmed by a hunter from Qaanaaq. He could only name two hunters that had done so recently and said that they had to take a route to the north via the Ikineq glacier in the inner part of Siorapaluup Kangerlua/Robertson Fjord. Another hunter from Qaanaaq spoke of this change in route as well, adding that it is no longer possible to drive up onto the Inland Ice Cap at Arfalluarfik/Diebitsch Glacier near Neqi (this includes two routes: one which goes directly up from Neqi and another which goes around Karrat/Kap Powell and then up Arfalluarfik/Diebitsch Glacier) – partly because there is no ice along that stretch of coast and partly because the gorge behind Neqi is no longer filled with snow and is also very steep. Five hunters from Savissivik mentioned that the decline in sea ice and poor ice conditions have meant that hunters no longer drive out on the offshore drift ice to hunt bears. One of them said that his last bear hunting trip on the drift ice had taken place in the early 1990s.

The majority of the responses to this question indicated that hunting patterns have changed because bears occur closer to the community (*i.e.* inhabited and areas with a regular human presence and hunting activity). Six of the responses did not include any detailed explanation of what has caused the change in distribution patterns, but one of the responses mentioned that bears come closer to the town of Qaanaaq. Two other respondents indicated that polar bears are now more frequently sighted in the period when there is open water. A hunter from Qaanaaq said that he no longer has to drive north to Ikersuaq/Kane Basin because he is now able to catch bears in the "local" area (during the interview he gave details of two polar bear catches in the vicinity of Kiatak/Northumberland Island in 2005).

However in eight cases, respondents told the interviewer that the change in distribution is due to changes in ice conditions, *i.e.* the decrease in the duration and amount of ice in the offshore areas and the resulting increase in the amount of open water (Table 29). One informant added that this may be the result of changes to current patterns coupled with the increase in temperature. A 47-year-old hunter from Qaanaaq town stated that there had not been many polar bears in the 1970s and that they now come closer and even come as far as Kangerluarsuk/Bowdoin Fjord just east of Qaanaaq, apparently because there is now more open water offshore (*i.e.* in Ikeq/Smith Sound). Another man from the same town was of the opinion that the bears gather on the remaining ice in the fjords where they seek food.

A 48-year-old from Savissivik made the following statement on the subject:

> I think the reason why the bears come closer and closer is that the sea does not freeze over any more during the winter time of year. Therefore, the bears come closer and closer. Since our sea has begun to freeze late, some bears have become very thin. When the ice forms early, the bears we catch are usually fat and taste good. Since the late 1990s, due to the fact that the sea freezes late, almost all of the bears that we have caught have not had any fat on them. The bears we caught this year were like that too. They have no blubber on them and they are not fat. If you disregard the *pingajoqqat* (mothers with two cubs; author's note) that we caught, which were a little bit plump … the rest of the bears that we catch are thin. That is because the sea out there does not freeze over – The difference is noticeable.

A 58-year-old from Siorapaluk said that more bears are now caught from boats in Ikeq / Smith Sound because there is less ice (see also the section "Hunting methods" under "The catch in figures", p. 187). A single hunter from Qaanaaq was also of the opinion that there had been a change to the hunt and said that the polar bears are more thin nowadays.

A 64-year-old hunter from Qaanaaq was of the opinion that the number of polar bears had increased since the introduction of regulations prohibiting hunting of bears from large boats and planes:

> No, I don't think they (climate changes; author's note) have noticeably affected it (the hunt; author's note), but I do think that when, for example, we see bears going through trash in Alaska – and not long ago there was a bear that almost ate someone – it makes you think about the time when planes were used to hunt bears, and they have stopped hunting like that now, or with big ships. Since they stopped that, I think there have been more and more bears. It's obvious.

These forms of hunting, which where practised in Alaska and around Svalbard were banned when the International Agreement on the Protection of Polar Bears and their Habitat was established (see *e.g.* Born & Rosing-Asvid 1989).

In Qaanaaq a 54-year-old hunter, who did not answer the question directly, philosophised about whether the bears lived healthier lives in earlier times because there were fewer of them:

> Now, since there are many of them, it obviously affects their search for food because they disturb each other. A bear looks for food in a very large area. Therefore if there are too many bears, they have to share a hunting area so that maybe only one of the bears gets enough to eat. I think that the bears lived a healthier life then, but now it is a pity for those that starve. It is probably those bears that come close to populated areas, I would say. The ice-covered area around us is not that big, apart from the great sea off Savissivik (*i.e.* in Baffin Bay; author's note). So if there are too many bears, they will naturally disturb each other.

See also the responses to question 26.

26. Have there been any changes to the polar bears that you have seen or caught?

In asking this question we were interested in the extent to which bears had become thinner, fatter, or changed in physical condition.

Of the 21 respondents to this question, 11 (52%) answered "yes", 8 (38%) answered "no" and 2 (10%) had no opinion on the subject (Table 30). There was however no statistical difference in the number of respondents who answered "yes" and "no" respectively ($\chi^2 = 0.244$, $P = 0.621$, $df = 1$).

All of the affirmative responses indicated that the bears are thinner. A 44-year-old informant from Siorapaluk stated that the bears were fatter in the late 1970s and 1980s, and that he now mostly catches lean bears. According to a 66-year-old hunter from Savissivik, the bears are thinner than they were in the past. In his opinion this is a result of the lack of snow on the ice and the late formation of the ice, which has meant that there are fewer ringed seals. Previously, when the ice formed in the autumn, a large number of seals were caught. That would also be the case now

	Total	Siora-paluk	Qaanaaq-Qeqertarsuaq	Savissivik
Yes	11	1	4	6
No	8	2	5	1
No opinion	2	-	1	1
N_{total} (hunters asked)	21	3	10	8

Table 30. Distribution of the responses to question 26, from the Qaanaaq municipality: "Have there been any changes to the polar bears that you have seen or caught?"

if the ice did not form just before the waves come and break up the ice causing the seals to disappear. According to the same man, there are now fewer, larger seals (this may be construed to mean that there has been a decrease in the number of young and subadult seals and that the remaining animals are adults; author's note). Conversely, a 48-year-old hunter from the same settlement indicated that the bears are thin because there are so many of them. He implied that the bears had increased so in number due to hunting regulations that they now compete for food and become thin:

> Like the others, I would say that the bears that we used to catch back in the days when we went out onto the open sea were very fat. Now the bears are getting thinner and thinner as more restrictions on them are introduced. The bears have begun to gather together so much that when one bear is sitting and waiting at a seal's breathing hole, another bear will come running over to it so that it runs away and neither of them will make the catch. They have started to run up to each other and disturb each other because they live in the same area [........]. When I was a child, bears never came visiting here. We did not even see bears when there was newly-formed ice. But now they come to visit regularly [........]. They can smell each other and locate each other and in that way, they disturb each other.

A 46-year-old from Savissivik told the interviewer that in recent years, he had noticed that the bears with darker fur (*i.e.* more yellow) are thinner than other bears.

Among the informants who did not think that there had been any change to the bears, there was however one hunter from Qaanaaq who added that he had heard that the bears that have been caught around Savissivik in recent years have been thin. Other statements made in response to this question were (1) that the bears that are skilled hunters are fat, whereas the less skilled bears are thin, and (2) that older bears are thin.

27. Have you observed changes to the occurrence of polar bears?

All of the 21 interviewees who were asked this question answered in the affirmative, indicating that they had all seen changes to the occurrence of polar bears. Several of the responses to this question coincided with the responses to the question about to what extent the physical factors have affected the bear hunt (question 25).

	Number of responses	Siora-paluk	Qaanaaq-Qeqertarsuaq	Savissivik
Closer	9	1	4	4
- due to changes in the ice	8	1	3	4
- because there are more of them	3	3	-	-
Other	1	1	-	-
N_{total} (responses)	21	6	7	8

Table 31. Distribution of the responses to question 27, from the Qaanaaq municipality: "Have you observed changes to the occurrence of polar bears?"

A total of 20 (95%) responses to this question included statements indicating that bears now come closer to the coast and areas that are regularly used by humans for travelling and hunting. Around 43% of these responses contained no additional explanation as to the reasons for this, while ca. 38% indicated that the differences are due to changes in ice conditions. Three (ca. 14%) hunters from Siorapaluk were of the opinion that the changes are the result of an overall increase in the number of bears. A single response (5%) included information on tracks (Table 31).

Of the nine hunters who said that the bears have generally come closer, one man from Siorapaluk specified that this also applies to the area around Nunarsuaq/Inglefield Land in southern Ikersuaq/Kane Basin. A hunter from Qaanaaq said it is very easy to come across bear tracks when hunting on the thin ice in the mouth of Murchison Sound and a 63-year-old informant from Savissivik said that the bears that occur along the coast are mostly young animals. That hunter suggested that the altered occurrence might partly be explained by the poor ice conditions and partly by the fact that the population of polar bears has increased:

> We have also noticed that the footprints of smaller bears have become very common, the tracks of smaller bears which are, for example, from last year. It is very common to see them now [........]. I don't understand why there are so many small bear tracks at this time, and they are also very thin. [........] It seems as though they have lost their mothers. It has been said, that the bears have started to die because there is no ice, that might be why. I think that might be why, maybe because their mother is dead, they are alone, or maybe they can't find their mother. I don't know. [........] like, for example, this year three bears, which were alone and not that big, have been caught [........] Maybe there are now so many bears that they disturb

> each other and spend time on doing other things instead of hunting, and therefore go hungry, so that they lack fat because they disturb each other. I suspect that this is the case.

Eight of the answers to this question indicated that the changes to ice conditions are the primary reason for the bears increased proximity to the coast. Several respondents suggested that the decrease in drift ice "out at sea" (*i.e.* Ikersuaq/Kane Basin, Ikeq/Smith Sound and the northern part of Baffin Bay) has meant that the polar bears are forced to come closer to the shore in order to seek out the areas of land fast ice in the fjords, according to two of the informants. A 48-year-old hunter from Savissivik said that changes to the distribution have been observed since the late 1990s when the bears began to get thinner. A 49-year-old hunter from Qaanaaq voiced however the opinion that the bears had begun to come closer to the shore during the 1980s. The same man also thought it likely that the polar bears prefer to hunt in areas of stable, snow-covered ice. He said that there has been a noticeable increase in the number of bears, which results in them being more and more dispersed. During this process, the bears have a tendency to stay close to each other because the bears "like to be not that far from each other".

Three hunters – all from Siorapaluk – were of the opinion that there had been an increase in the number of polar bears (Table 31). One of them (aged 46) said that it had been difficult to catch polar bears in the 1970s and 1980s, but that during the period 1980–1990 the number of bears had slowly increased. However, although he would not preclude that the increased occurrence might be because "there is more sea" (*i.e.* a decrease in sea ice), he held instead the perception that it was due to an increase in the population. A 64-year-old hunter from the settlement of Qeqertarsuaq recalled that when he was 18 years old (*i.e.* in the 1960s), there were many polar bears around Umimaat Nunaat/Ellesmere Island in Canada and very few in Greenland (*i.e.* in the north of Ikeq/Smith Sound and Ikersuaq/Kane Basin; author's note). According to the same man,

> Nowadays, there are a lot of bears there (northern Ikeq/Smith Sound; author's note) and it is as though they come over from Canada. We can see that the tracks go that way, even though the ice is breaking up. It sometimes breaks up with a few days' interval. The tracks come this way in the spring.

A 43-year-old from Qaanaaq said that in the 1970s and 1980s, when the hunters travelled north to Nuussuaq/Washington Land, there had not been many bears in the eastern part Ikersuaq/Kane Basin. He thought that more bears had come in the mid-1980s and that even more have come since then.

One of the responses to this question (category: "Other"; Table 31) included information indicating that there are now more tracks near Nunarsuaq/Inglefield Land.

The biology and behaviour of the bears

28. Have you seen mating, mating behaviour or any tracks that indicate these behaviours?

Of the 23 who were asked this question, 15 (65%) answered "yes" and 8 (35%) answered "no" (Table 32).

	Total	Siora-paluk	Qaanaaq-Qeqertarsuaq	Savissivik
Yes	15	6	6	3
No	8	-	5	3
N_{total} (hunters asked)	23	6	11	6

Table 32. Distribution of the responses to question 28, from the Qaanaaq municipality: "Have you seen mating, mating behaviour or any tracks that indicate these behaviours?"

None of the informants had actually observed mating polar bears, and their responses were mostly concerned with observations of tracks which indicated mating activities (Fig. 18). The statements regarding the mating season indicated that it takes place between March and May, peaking in April. The data (including the information that was drawn on maps) given in response to the question of the "month in which the mating season takes place" (n = 16) was distributed as follows: March (3 responses), April (10) and May (3).

	Total	Siora-paluk	Qaanaaq-Qeqertarsuaq	Savissivik
Yes	22	6	10	6
No opinion	3	-	1	2
N_{total} (hunters asked)	25	6	11	8

Table 33. Distribution of the responses to question 29, from the Qaanaaq municipality: "What do the bears eat, apart from ringed seals?"

29. What do the bears eat, apart from ringed seals?

Of the 25 respondents to this question 22 (88%) talked about polar bears eating other animals as well as ringed seals, while 3 (12%) had only seen ringed seals in polar bear stomachs (Table 33).

Several different food items were mentioned. About 44% of the answers cited grass, plants, mosses and seaweed (Table 34). Two hunters from Qaanaaq stated that the bears may eat grass in the spring, summer and winter. An informant from the town spoke of how the thin bears have seaweed and other things in their stomachs, as they can eat anything. He once caught a lean bear while it was eating seaweed by Upernavissuaq on Qeqertarsuaq/Herbert Island. Another hunter had

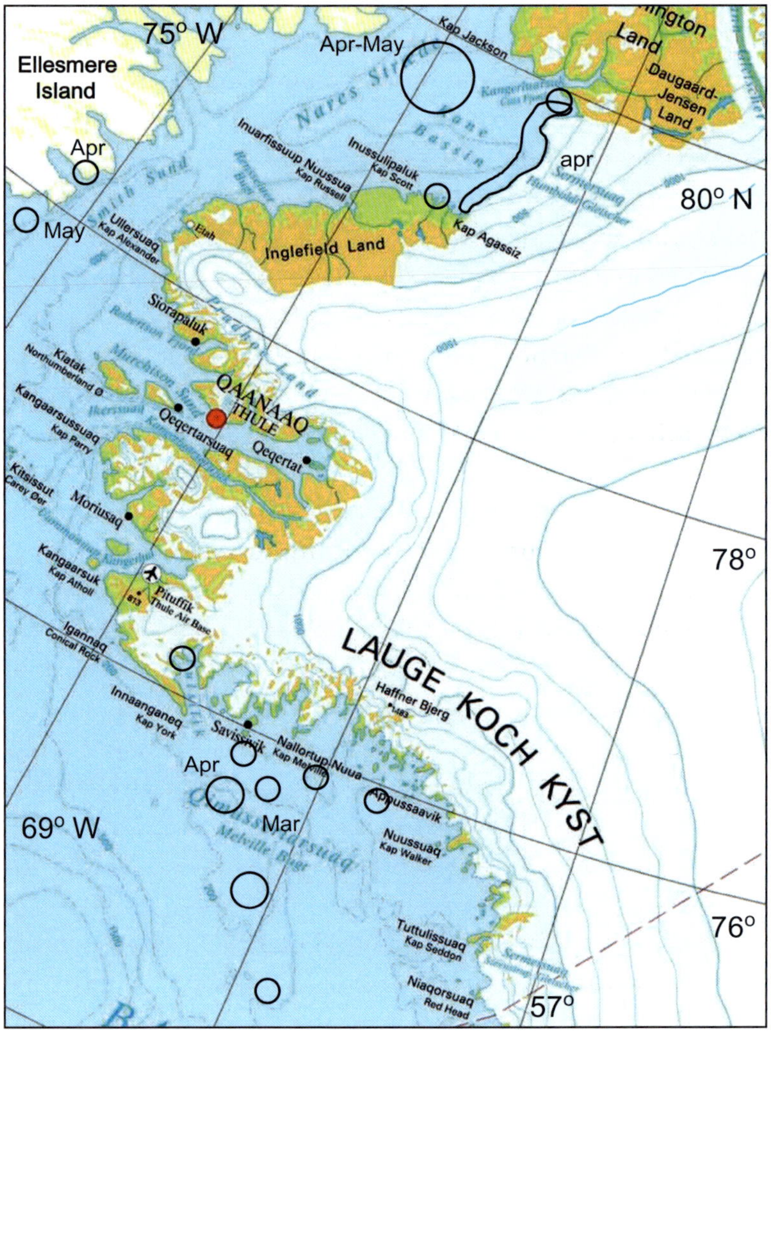

FIG. 18.
Locations where polar bear mating behaviour or tracks indicating mating behaviour have been observed, as drawn on a map by 14 hunters during the interviews in the Qaanaaq municipality in February 2006.

	Number of responses	Siora-paluk	Qaanaaq-Qeqertarsuaq	Savissivik
Grass, plants, moss	11	2	4	5
Seaweed	3	1	2	-
Other seals and walruses	4	-	4	-
Whales	2	-	2	-
Bear cubs	3	1	1	1
Other (*e.g.* birds, sharks, meat caches)	9	4	3	2
N_{total} (responses)	32	8	16	8

Table 34. Types of responses to question 29 in the Qaanaaq municipality: "What do the bears eat, apart from ringed seals?"

shot a bear with seaweed in its stomach, in the vicinity of Kangaarsussuaq/Kap Parry. A 66-year-old hunter from Savissivik had seen bears that had eaten grass, moss and mountain crowberry (*Empetrum hermaphroditum*). Two of the interviewees from Savissivik were of the opinion that the phenomenon of bears eating grass has become more frequent in recent years (for example seen at Innaanganeq/Kap York), and one of these informants suggested the explanation that the bears might go inland in order to eat grass when they get hungry.

Bearded seals and walruses were also specified as the polar bears' prey. A hunter from Qaanaaq gave an account of a bear whose stomach contained both narwhal and bearded seal remains, while a 64-year-old informant from the same town told of how he had seen a male walrus lying dead on the ice by an iceberg in the vicinity of Umimaat Nunaat/Ellesmere Island, in 1966. The walrus had been killed during a fight with a bear, which had however not eaten its prey. He also said that other hunters had told him about the case of a bear that "was not that big" had killed a pregnant walrus at its breathing hole (see p. 88). The same man also recounted the tale of an unusual case where a walrus appeared to have eaten a polar bear:

> We came up on the leeside of a walrus, which had come up through a crack in the ice and we caught it – it was in Ikeq/Smith Sound. It had bear fur in its stomach, so we could tell that it had eaten bear. How else could it have got bear fur in its stomach?

Two hunters from Qaanaaq told of how, a few years earlier at Kiatak/Northumberland Island, they had shot a bear which had narwhal mattak (skin) in its stomach, presumably from a small whale.

Some of the informants mentioned polar bear infanticide and cannibalism (Table 34), and that adult males (*"ittorruit"*) in particular hunt and eat cubs. One hunter said:

> Once there was someone – I can't remember who it was – who caught a bear in Ikeq/Smith Sound with a bear cub in its stomach. That's how these animals are, they love nothing better than to eat bear cubs. It is the large adult bears that behave like that.

Another informant said:

> There was a case where a large male bear killed a sleeping bear and then ate it but didn't eat it all up. I know that if a large old male bear comes across the tracks of *ateqqaat* (cubs that have just left the den; author's note), it will pursue them for a long time, where they once in a while have to lie down. This is in order to eat them.

Other animals eaten by the polar bears included Brünnich's guillemot (*Uria lomvia*) in June at Moriusaq, Greenland shark carrion (*Somniosus microcephalus*) in areas where these are caught, and sled dogs. Five informants spoke of how polar bears are able to raid meat depots to find food. Furthermore, a 54-year-old hunter from Qaanaaq told of an unusual occurrence. When asked if polar bears eat anything other than ringed seals, he responded by saying:

> Yes, hair. We were on the way to Savissivik because my late name-brother "NN", "NN" and I were going to Kullorsuaq. We drove out quite a way. I think it (the bear; author's note) had eaten hair which had fallen out due to brushing, because it was ravenous. We caught it, and it had human hair in its stomach. It was very hungry and it can eat anything when it is hungry.

30. Have you seen polar bears hunting?

Of the 23 informants who were asked this question, 11 (48%) answered "yes" while the remainder answered "no" (Table 35). The majority of the responses included observations of polar bears hunting ringed seals either at the seals' breathing holes, in their birth lairs, or as *uuttoq* (*i.e.* seal that is basking on the ice). A single response included information on polar bears hunting walrus and bearded seals (Table 36).

A hunter from Siorapaluk had observed bears sitting by breathing holes waiting for ringed seals, and had seen evidence suggesting that bears are able to dive down into or resurface through breathing holes as small as 50 cm in diameter. Another observation made by a hunter from Qaanaaq confirms just how supple polar bears are:

> An adult male bear had followed a frozen lead in the ice. When it came across a seal breathing hole, it had waited. But after a while, it began to open the breathing hole, made an opening the size of its head, and went down into the water through it. I wondered how it had been possible and followed the frozen crack. The distance was the same as from here (*i.e.* the municipal office in the town of Qaanaaq; author's note) to the power plant (circa 100 m; author's note) – the distance to the next breathing hole – when it reached it, it came up through it with ease and continued on its way. That was what puzzled me. Maybe it can do something with its pelvis so that it can get through, that's how supple it is.

	Total	Siora-paluk	Qaanaaq-Qeqertarsuaq	Savissivik
Yes	11	2	4	5
No	12	2	7	3
N_{total} (hunters asked)	23	4	11	8

Table 35. Distribution of the responses to question 30, from the Qaanaaq municipality: "Have you seen polar bears hunting?"

	Number of responses	Siora-paluk	Qaanaaq-Qeqertarsuaq	Savissivik
Ringed seals				
- at breathing holes	8	1	4	3
- in lairs	4	1	-	3
- as *uuttoq*	1	-	-	1
Other	1	-	1	-
N_{total} (responses)	14	2	5	7

Table 36. Types of responses to question 30 in the Qaanaaq municipality: "Have you seen polar bears hunting?"

Savissivik in May 1992. Photo: E.W. Born

Two hunters from Qaanaaq had also seen bears sitting and waiting by a seal's breathing hole. One of these informant's observation was made in an area just to the south of Nuussuaq/Washington Land. One of the three informants from Savissivik (who had seen such "still hunting" at a breathing hole) mentioned how he had observed a couple of bear cubs participating in the hunt in the vicinity of Salleq/Bushnan Island. The cubs seemed to be helping their mother by causing a disturbance at other breathing holes so that the seal would only surface through the hole where their mother sat waiting. A hunter from Qaanaaq spoke of a bear in the Nuussuaq/Washington Land area that sat by a breathing hole for such a long time that in the end it had created a very deep indentation in the snow. The bear had scratched the surface of the ice covering the breathing hole until it was so thin that the seal could easily be snapped up when it came up for air. He had seen this phenomenon several times.

Some hunters had also observed bears hunting ringed seal pups in their lairs. A hunter from Siorapaluk spoke of how in the spring, the hunters follow polar bears which are hunting ringed seals in their birth lairs:

> It is by following the bears' track, that we hunt polar bears. We call what they do in the month of April "*natsiarriortut*" – "their seal pup hunting season".

He said that when the sun returns in late February, the polar bears start hunting ringed seal pups in their birth lairs and that he had once during a hunting trip come across a bear that was destroying a birth lair which was hidden in the snow. The bear destroyed the lair in a violent fashion causing a flurry of snow. A hunter from Savissivik, who had observed polar bears in several locations hunting ringed seals, mentioned how bears are able to drag seals out of their lairs even when they are deep inside the lair. He spoke of a case in which a bear had dragged a dead seal out of its lair and laid it out to cool, after which the bear fled the hunters. The bear had bitten the seal's skull to shreds. Another hunter from Savissivik spoke about an observation from Qeqertaq/Salve Island where a couple of bears wandered around the icebergs destroying the ringed seals' birth lairs in order to find food. A third informant from the same settlement reported a couple of observations of the bears' seal hunt. In April 1970, he had seen a female bear with two cubs, which:

> were really out to get something to eat. They crawled in between some very tall ice hummocks. As it turned out, the mother was eating seal pups with a yellowish-white pelt, young seals that had not grown vibrissae yet. The female bear was very lean. In fact, when they are very hungry the bears are able to eat yellowish white seal pups without allowing them to cool down first. However, I have seen several that were not hungry kill these seal pups, drag them out and just leave them there. Once when we drove alone to the south towards Kullorsuaq [........], we arrived at a place to the north of Qapiarfissalik. There, we followed a bear for quite some time and it dragged a whole lot of seal pups up. It just left them lying there, because they had no blubber on them.

A hunter from Savissivik had seen a bear which ran from *uuttoq* to *uuttoq*, until it finally caught a seal. Subsequently it cooled the prey down by covering it with snow. Furthermore, the bear used the seal as a kind of pillow until it had cooled down. He had also seen another bear which had covered a seal in snow and used it as a pillow while it waited for the seal to cool down. Several times, he had seen bears which allowed their prey to cool down before eating it.

A hunter from Qaanaaq gave a detailed account of the observations he had made of bears' hunting behaviours. He said that there are a large number of bearded seals in the Nuussuaq/Washington Land region and the bears catch a lot of bearded seals in that area:

> When we happen to disturb a bear that has just caught an adult bearded seal, it may grab the bearded seal between its teeth and run away with it (an adult bearded seal weighs 250–300 kg; Burns 1981).

The same man reported having seen a bear that had killed a walrus using a lump of ice:

> We went out in February to celebrate that the sun had returned and caught 4 bears [........]. I had just left Anoritooq/Kap Inglefield, went south past Qamaaqvik and was about to arrive at Naujaat when I saw two walruses on the ice. I stopped and waited for the others. When they got to me, I said that I wanted to get some food for the dogs that was not frozen and that I would attempt to take the smallest walrus. I already had enough food for the dogs though. Then I began to approach them on the ice and once I got closer, I could see that the scene looked a bit odd – there was blood on the ice. They turned out to be some walruses that had recently been killed by a male polar bear. I looked around for its tracks and it was very cold, the crack in the ice was not that far away from me. I put my finger on its urine, and my finger went right through – it was not yet frozen. When the bear had detected that there was someone close by, it urinated. The bear had caught a walrus. Before it had sat down to wait for the walrus, it had fetched a large piece of salt water ice from the tidal zone and had worked it so that it was smooth all over. It had fashioned a tool with which it could hit the walrus on the head. It was completely smooth all over, with no rough spots at all. Then it sat down behind some ice hummocks, and when the walrus came up for air, the bear pounced over to it. Its claws had made deep impressions in the new ice [........]. It had attacked the walrus and started to hit it over the head with its weapon. It was an old, pregnant walrus which had very fine tusks. Now, a walrus is very supple when it is in the water because it has large fore flippers, but the bear had hit it over the head and had smashed its skull from just over its upper lip all the way to the back of its head. Its skin in that area was shredded, that's how hard it had hit it on the head. When it had hauled it up out of the water, it had dragged it a short distance from the walruses' breathing hole and then repositioned it where it had dragged it out of the water. It was so recently caught that it was not yet frozen so I was able to simply cut off its head. After the bear had dragged its prey around, it had put it back by the hole and had placed its club beside the dead walrus. Then it had pulled its foetus out, eaten the stomach of the foetus and then left it to cool down so that it could later eat its fill of the foetus which is its favourite food. However, it did not get any more of it because I took it with me on the sled since it was not frozen.

Photo: E.W. Born

31. Have there been any changes in what the polar bears eat?

This question was put to 16 hunters, of whom 2 (13%) answered "yes" and 14 (ca. 87%) said "no" (Table 37).

One of the two informants who answered in the affirmative said that the polar bears now seek out meat depots more frequently than they used to, while the other told of how polar bears go inland to eat grass more frequently nowadays. One of the respondents who said "no" said that bears also ate grass in the past.

	Total	Siora-paluk	Qaanaaq-Qeqertarsuaq	Savissivik
Yes	2	-	-	2
No	14	4	5	5
N_{total} (hunters asked)	16	4	5	7

Table 37. Types of responses to question 31 in the Qaanaaq municipality: "Have there been any changes in what the polar bears eat?"

The Upernavik municipality

Area 3 (Kullorsuaq and Nuussuaq)

We interviewed a total of 18 hunters in the settlements of Kullorsuaq and Nuussuaq (Table 1). However, in the case of one of the interviews, the recording equipment malfunctioned, which resulted in the highest possible number of respondents from this area being 17.

The answers given to questions 1 and 2 ("How many bears do you think were caught in your town/settlement in 2005?" and "How many bears do you think have been caught in the municipality in total?") are summarised in the section "Number of animals taken" in "The catch in figures", p. 193.

Hunting and travelling conditions

3. Do you catch more bears than you used to?

Six (40%) of the 15 respondents to this question answered "yes", 4 (ca. 27%) answered "no", 1 (ca. 7%) was of the opinion that the catch had decreased, while 4 (ca. 27%) did not express any particular opinion on the subject (Table 38).

Of three hunters from Kullorsuaq who reported an increased catch, one 44-year-old hunter stated that there were very few bears in the 1970s and that back then catching a bear had only been possible after a lengthy search, whereas during the 1980s bears could be caught after just two or three days of hunting. This hunter had a theory about the relationship between bears and ringed seals:

> .... when the number of bears increased in the 1980s, the seals ... we call seal pups aninerit ("those which have come out"; *i.e.* ringed seal pups that still have their lanugo coat, author's note) ... when they start to give birth to pups in April, the bears start to eat them. Back then, all the small ringed seal young came to the coast, when the ice came [........] during the 1970s and 1980s. But according to my estimations the seals have been bigger, in recent years. At the start of the 21st century, or at the end of the 1990s, we started to catch bigger ringed seals, which we hadn't otherwise been used to catching in the past. As the bears

	Total	Kullorsuaq	Nuussuaq
Yes	6	4	2
No/Same	4	3	1
Decreased	1	1	-
No opinion	4	1	3
N_{total} (hunters asked)	15	9	6

Table 38. Distribution of the responses to question 3, from Kullorsuaq and Nuussuaq (Upernavik municipality): "Do you catch more bears than you used to?"

> increased in number, they started to eat seal pups, and perhaps ate them all. Therefore the ringed seal pups that otherwise should have reached the coast had been eaten, so more of the large seals came to the coast.

For further information on the occurrence of "large" seals, see also the section containing data from the Qaanaaq municipality in response to the question "Have there been any changes to polar bears that you have seen or caught?", p. 77, and from Area 3 (Nuussuaq and Kullorsuaq) under the question "Have you observed any other changes?", p. 119.

A 46-year-old hunter from Kullorsuaq was of the opinion that the catch had increased since the early 1980s, and that his own catch of polar bears had stabilised in the late 1980s because the number of bears had increased. In that period he shot an average of about six bears per year. A third informant from the same settlement said that the increase in the catch was due to the fact that the bears had come closer to the coast. This statement was supported by a 67-year-old hunter from Nuussuaq, who said that bears are now more frequently encountered, presumably because they have increased in number. A 42-year-old from this settlement was also of the opinion that the population had increased and that the polar bears also occur more frequently along the coast because of the decline of the sea ice.

One hunter from Kullorsuaq said that his catch of polar bears had decreased. However, this was because in the past he used to think that it was wonderful to go hunting, whereas in recent years he had begun to spend more time on fishing. A resident of the same settlement explained that his hunting activity is in part dictated by when, and to what extent, the purchasing of Greenland halibut stops. In 2005, for example, he went bear hunting when the fish purchasing stopped in April.

4. Are more bears caught here in the settlement than previously?

Among the 16 hunters who were asked this question, 10 (63%) answered in the affirmative, while 6 (37%) answered "no" (Table 39).

	Total	Kullorsuaq	Nuussuaq
Yes	10	7	3
No/same	6	3	3
N_{total} (hunters asked)	16	10	6

Table 39. Distribution of the responses to question 4, from Kullorsuaq and Nuussuaq (Upernavik municipality): "Are more bears caught here in the settlement than previously?"

Four hunters (1 from Kullorsuaq and 3 from Nuussuaq) thought that the increase could partly be explained by the fact that bears now occur closer to the shore (2 hunters) and partly by an increase in the number of bears (2 hunters). A 66-year-old informant from Nuussuaq expressed the latter as follows:

> Are there ever! A great deal more. During the 1960s [........] when we were hunting bears, we sometimes came across bear tracks and on occasion followed the tracks for one or two days before we caught the bear. Nowadays – this year in December there were bears here, underneath the KNI building (*i.e.* the shop) – in the dark. Fortunately, it wasn't caught. The cubs were very young too. Maybe they went down there, out to sea.

Some hunters did however indicate that the increased catch may also reflect an increase in hunting efforts. One said that skiffs, in contrast to the past, are also used as a vehicle in bear hunting. Another informant said that the human population has increased. A third informant stated that the polar bears are not hunted during the intensive hunt on small cetaceans during the summer (the man talked about *"qilaluaq"* without specifying whether it was *"qilaluaq qaqortoq"* or *"qilaluaq qernertaq"*, *i.e.* white or black, meaning beluga or narwhal; it is likely narwhal in Qimusseriarssuaq; author's note).

5. Do you use regular routes when hunting polar bears?

Of the 16 interviewees to whom this question was asked, 11 (69%) answered "yes" and 5 (31%) answered "no" (Table 40).

The hunters from Kullorsuaq told the interviewer how their routes tend to go west onto the drift ice or north or northwest in Qimusseriarsuaq/Melville Bay (Fig. 19). After the sea ice has become stable at some point in January, one can for example choose to travel to the west over the drift ice until April–May, and on occasion June.

	Total	Kullorsuaq	Nuussuaq
Yes	11	7	4
No	5	3	2
N_{total} (hunters asked)	16	10	6

Table 40. Distribution of the responses to question 5, from Kullorsuaq and Nuussuaq (Upernavik municipality): "Do you use regular routes when hunting polar bears?"

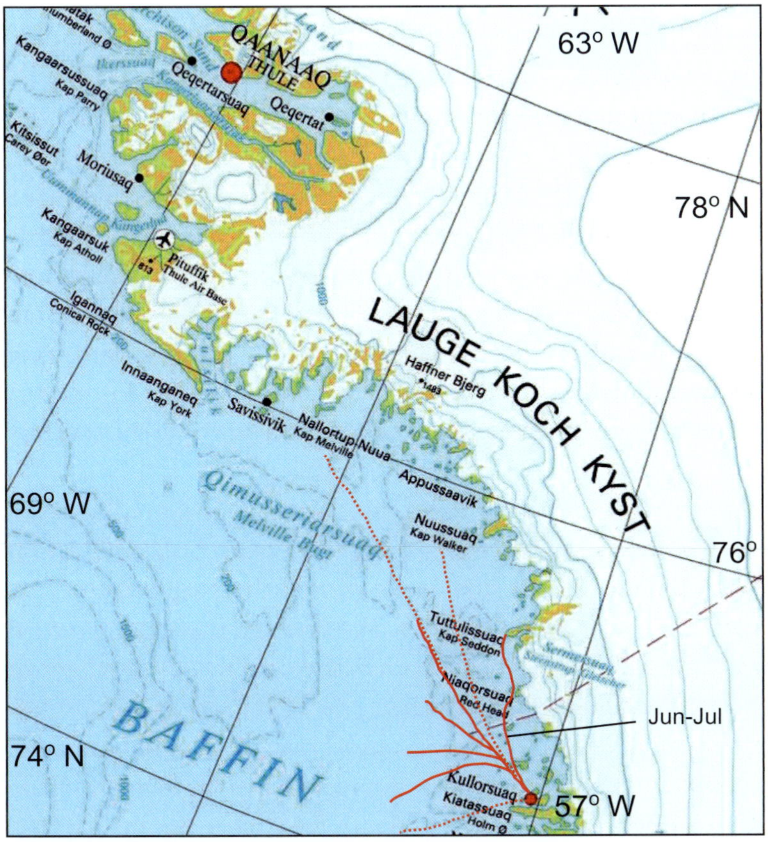

FIG. 19. Sled (solid line) and boat (dotted line) routes in Qimmusseriarsuaq/Melville Bay (Area 3) as drawn by five hunters during the interviews in Kullorsuaq and Nuussuaq in February 2006. The same routes are often followed whether the mode of transport is a boat or a sled. See also Fig. 24.

The location of the routes to the north or northwest partly depends on driving conditions. From some time in February, the route often runs along the edge of the land fast ice. Travellers often go to the north towards an area of shallow water where there are stranded icebergs ("*Qorfiit*") in Qimusseriarsuaq (Fig. 2). Two hunters said that they often reach the islands of Saattut/Sabine Islands in Qimusseriarsuaq during their bear hunts. A 63-year-old mentioned that it is now seldom that hunters go as far north as Savissivik. In addition, some informants told of how hunters travel to the north to Qimusseriarsuaq/Melville Bay in order to hunt narwhals. During these hunting trips, polar bears are also taken (see also question 8).

Generally speaking, the hunters from Nuussuaq use the same routes. A few hunters from this settlement said that they seek out areas of shallow water with stranded icebergs, both offshore and to the northwest. One of them told the interviewer that he goes to Tuttulissuaq/Kap Seddon in Qimusseriarsuaq, an area where a large number of bears have been caught.

6. Do a greater number of polar bears come to visit/come of their own accord?

Of the 17 respondents, there were 5 (29%) who answered "yes", 8 (47%) who answered "no", 1 (6%) who said that there are fewer visiting bears and 3 (18%) who had no opinion on the matter (Table 41).

A 47-year-old from Kullorsuaq said that the number of visiting bears had increased since he was a child, while a 63-year-old from the same settlement thought that the change had taken place since he was young. These statements indicate that there has been an overall increase in the number of "bear visits" since the 1960s. Two informants from Nuussuaq had also noticed an increase in the number of visiting bears, and one of them mentioned that in the winter of 2005/2006 there had been three instances of bears visiting the area around the local shop (see the answer to question 4).

However, the majority of the respondents did not believe that there had been an increase in the number of bears which come "on their own accord" and several respondents indicated that the number of such visits varies from year to year.

Two hunters from Kullorsuaq were however of the opinion that the increased boat and snowmobile traffic has pretty much scared the bears away so they no longer come close to populated and trafficked areas.

Data pertaining to individual bear catches indicates however that only a small number of polar bears come of their own accord to populated areas in this part of the Upernavik municipality (see "Hunting methods" under "The catch in figures" p. 196).

7. Have your routes changed in recent years?

	Total	Kullorsuaq	Nuussuaq
Yes	5	3	2
No/Same	8	5	3
Decreased	1	-	1
No opinion	3	3	-
N_{total} (hunters asked)	17	11	6

Table 41. Distribution of the responses to question 6, from Kullorsuaq and Nuussuaq (Upernavik municipality): "Do a greater number of polar bears come to visit/come of their own accord?"

A total of 15 interviewees were asked this question, of which 3 (ca. 20%) answered "yes" and the remaining 12 answered "no" (Table 42). The three informants who answered in the affirmative were 42, 66, and 66 years old, respectively, while those who said "no" had an average age of 47 years (sd =11.2; min. – max.: 34–67 years).

	Total	Kullorsuaq	Nuussuaq
Yes	3	-	3
No/same	12	9	3
N_{total} (hunters asked)	15	9	6

Table 42. Distribution of the responses to question 7, from Kullorsuaq and Nuussuaq (Upernavik municipality): "Have your routes changed in recent years?"

It is therefore difficult to say to what extent the response given to this question depends on age (and experience).

Most of the hunters were not of the opinion that the routes had changed to any great extent and some stated that their choice of route depends on the ice conditions and the occurrence of polar bears, as well as the preferences of the individual in question. It was however mentioned during the interviews that hunters seldom go as far as Savissivik nowadays and that it seems as though the bears occur closer to populated areas.

Three hunters from Nuussuaq did however say that the routes had changed. Two of these informants stated that hunters have not travelled to the west over the drift ice in recent years because the leads in the sea ice have been situated relatively close to the shore and one of them added that: "Sometimes, the bears land here en masse." The third hunter spoke of how the increase in the strength of the current due to the receding glaciers means that there are areas which are no longer passable by dogsled.

8. Are a greater number of polar bears caught from boats now than in the past?

Among the 17 polar bear hunters who were asked this question, 14 (ca. 82%) answered "yes", 2 (12%) answered "no" and 1 (6%) had no opinion on this subject (Table 43).

Several informants from both Kullorsuaq and Nuussuaq remarked that a greater number of bears are now caught from boats, partly because the boats are fast and

	Total	Kullorsuaq	Nuussuaq
Yes	14	9	5
No/Same	2	1	1
No opinion	1	1	-
N_{total} (hunters asked)	17	11	6

Table 43. Distribution of the responses to question 8, from Kullorsuaq and Nuussuaq (Upernavik municipality): "Are a greater number of polar bears caught from boats now than in the past"

partly due to the lighter ice conditions in the spring which allow them to go to the west to the drift ice in order to hunt bears.

An informant from Kullorsuaq said that more bears had been taken from cutters and skiffs then from dogsleds and that he disapproved of this development. Another hunter from Kullorsuaq expressed the same opinion:

> About that, I would say that there are more now (*i.e.* bears caught from cutters; author's note). This is because cutters have been sailing to the west of us more frequently in recent years. When narwhal licences are granted in the spring, and when the hunters come across bears while they are out hunting narwhals, the bears are usually caught. I mostly criticise the fact that the cutters sail to the west of us. I think that the cutters are intended for fishing and therefore I am usually of the opinion that they should be used for fishing to a greater extent. But in the spring, when the sea opens up and the cutters sail to the west of us, we come across them when we travel by dogsled. Being me is not much fun. For example, when we are searching for animals to hunt, I feel that the cutters disturb me when they come across me [........] when we dogsled drivers are out on a hunting trip. Therefore, I can say that there has been an increase. We hear about them too. That more bears have been caught from cutters in recent years. It's not just an isolated case ... one single cutter's bear catch. For example, this summer I saw a cutter from the Uummannaq area which had caught two bears. It landed here in the summer. That is how I have noticed that cutters, not just from the Upernavik municipality, are out hunting bears or something like that [........] When we are only out driving dogsleds, it is unpleasant when the cutters sail out to the west like that.

Another informant made similar comments and he, along with two others from Kullorsuaq and one from Nuussuaq, also mentioned that in recent years polar bears have been caught during narwhal hunts in Qimusseriarsuaq in June and July.

A couple of informants were of the opinion that a large number of polar bears have been taken from boats in recent years in the area to the south of Nuussuaq.

The information regarding the individual bear catches indicated that there has undoubtedly been an increase in the percentage of bears that are caught from boats in this part of the Upernavik municipality (see "Hunting methods" under "The catch in figures", p. 196).

9. Are a greater number of polar bears caught on land than in the past?

This question was put to a total of 17 informants. One (6%) answered "yes", 10 (59%) answered "no" and 6 (35%) had no particular opinion on the subject (Table 44).

The informant who answered "yes" gave a response in which he reasoned that a greater number of bears must be caught on land since, in his opinion, a greater number of bears are caught in the summer on the whole.

	Total	Kullorsuaq	Nuussuaq
Yes	1	1	-
No/Same	10	5	5
No opinion	6	5	1
N_{total} (hunters asked)	17	11	6

Table 44. Distribution of the responses to question 9, from Kullorsuaq and Nuussuaq (Upernavik municipality): "Are a greater number of polar bears caught on land than in the past?"

Occurrence of polar bears and dens

10. Do the bears occur in particular areas?

Of the 16 informants who were asked this question, 8 (50%) answered "yes", 7 (44%) "no" and 1 (6%) had no particular opinion on the matter (Table 45).

	Total	Kullorsuaq	Nuussuaq
Yes	8	5	3
No	7	4	3
No opinion	1	1	-
N_{total} (hunters asked)	16	10	6

Table 45. Distribution of the responses to question 10, from Kullorsuaq and Nuussuaq (Upernavik municipality): "Do the bears occur in particular areas?"

According to the responses given, the bears occur (1) out on the drift ice and on the land fast ice in Qimusseriarsuaq/Melville Bay and (2) at the leads which run parallel to the shore (*i.e.* in the shear zone between the land fast ice and the drift ice); Table 46; Fig. 20).

FIG. 20.
The general winter and spring distribution of the polar bears in Qimmusseriarsuaq/ Melville Bay (Area 3) as drawn on a map by five hunters during the interviews in Kullorsuaq and Nuussuaq in February 2006.

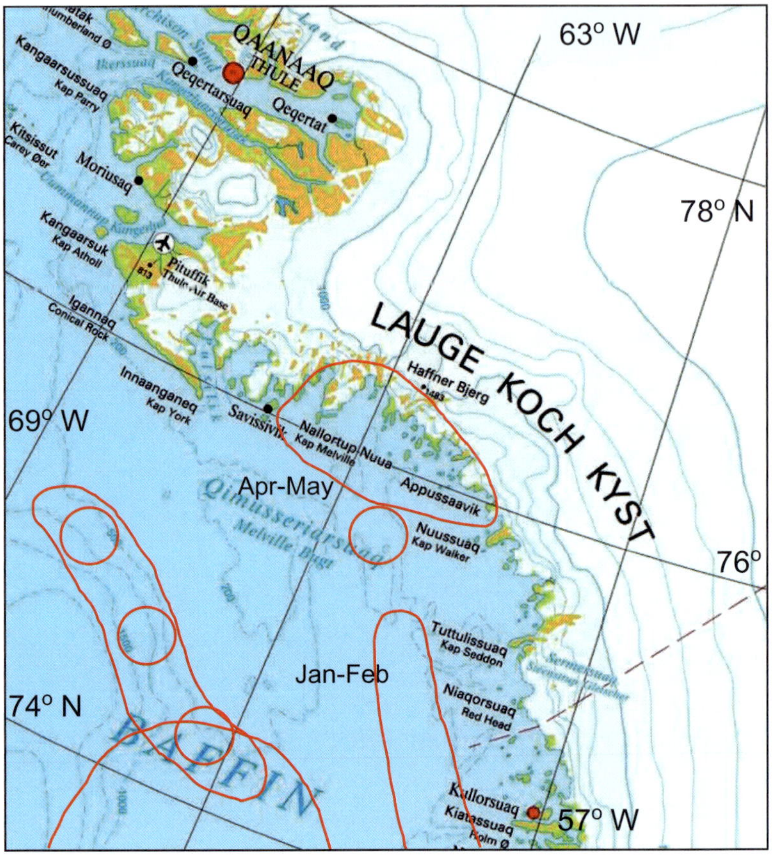

A hunter from Kullorsuaq stated that the polar bears inhabit the ice around Saattut/Sabine Islands in Qimusseriarsuaq in the spring because that is where the ringed seal pups (*aninerit*) are to be found. Another informant from the same settlement said that bears occur mostly to the west during the winter "at about this time" (February; author's note) whereas in the spring they move towards the

	Number of responses	Kullorsuaq	Nuussuaq
Drift ice to the west	6	4	2
Leads in the ice	1	1	-
Other places	1	-	1
N_{total} (responses)	8	5	3

Table 46. Types of responses to question 10, from Kullorsuaq and Nuussuaq (Upernavik municipality): "Do the bears occur in particular areas?"

glaciers where they remain during the month of May. Another resident of this settlement was of the opinion that the bears seek colder areas in the summer and autumn but did not specify where these areas are. A response given by a 50-year-old hunter from Kullorsuaq indicated that the distribution of polar bear depends on ice conditions and current patterns:

> Bears occur in specific areas ... as I have already said, right? When there is sea ice in the winter, they head east when the current is less strong. Then when the current gets stronger, they head to the west. The areas that they seek out ... these places, they all have many icebergs. Many icebergs frozen in place in shallow water. These are the places that the bears head for, you know. And the areas at the edge of the glaciers ... when they have been there, then they head west [........] The routes taken by the bears, when they go west are always the same, you know. In the area close to us, just to the west.

A 46-year-old hunter from Kullorsuaq gave an answer which indicated that the occurrence of bears has changed:

> In December they come closer to here. When the ice has just formed here and when the sea has just frozen over they come closer to this area. Then this area is full of bears [........] Sometimes they come in here, around the area close to us. [........] In December the bears seek food constantly, because they are hungry. Then they are no longer afraid. They are not scared of anything anymore. This is the month when it gets most hungry, this animal [........] In the 1970s I didn't see as many bears here, in this area [........] I don't know how the polar bear's situation was back then; whether they were just further away or if there were fewer of them. I have no idea what to say about that. Were there fewer back then, or were they just a long way away from here? The bear seeks food. When there are many seals, then there are many bears [........] Right now the ice to the north of us varies greatly in quality [........] In January–February we get open water all the time. The ice is continually destroyed. Then when the new ice comes, we hear that the bears are in that area. They always behave in that way. Naturally, when we have open water and when it is impossible to travel on the ice, the bears do not frequent those areas.

A hunter in Nuussuaq recounted how bears most often occur on the offshore drift ice where there is food but that they travel to the east when the ice consolidates. When the ice disappears again, they move once more to the west. He added however that bears can occur all along the coast and that here they are very difficult to encounter because they are skilled at hiding as soon as they hear a sound.

Two hunters from Kullorsuaq and Nuussuaq stated that the bears prefer to stay at the leads in the ice which run along the shoreline. One of these informants, a 66-year-old resident of Nuussuaq, explained how the cracks form in the ice when the current grows stronger because of a waning moon ("... when the moon disappears"), and how at that time, the bears head towards the leads. When there is a new moon ("... it comes back again, very thin"), the bears move towards the east in large numbers. The perception that the bears' movements are to some extent dictated by the lunar cycle and the condition of the tidal current was also mentioned by another hunter from Nuussuaq.

A shallow water bank to the southwest of Nuussuaq, Kitsissorsuit / The Ederfugle Islands and the area west of Kjers Glacier (just south of Tuttulissuaq / Kap Seddon) were cited as being "bear areas"; but other areas were also indicated (Fig. 20).

Those informants who were not of the opinion that the bears occur in particular areas said that the polar bears are constantly on the move and can occur anywhere depending on ice conditions and where they are able to find food.

11. Where have you seen bear tracks?

All of the 16 respondents to this question told the interviewer where tracks can be observed. Moreover, four hunters showed on maps some areas where tracks are frequently observed (Fig. 21).

Some responses to this question indicated that tracks are observed in particular on the western drift ice while others indicated that they have come closer to populated areas. Other informants were of the opinion that tracks can be observed everywhere (Table 47).

One respondent said that tracks are found in the areas which bears are believed to frequent:

> When we are out hunting seals on the ice we travel to specific places, when we begin to get a sense of where the bears are heading. In the direction of Niaqorsuaq – on occasion towards Tuttulissuaq / Kap Seddon.

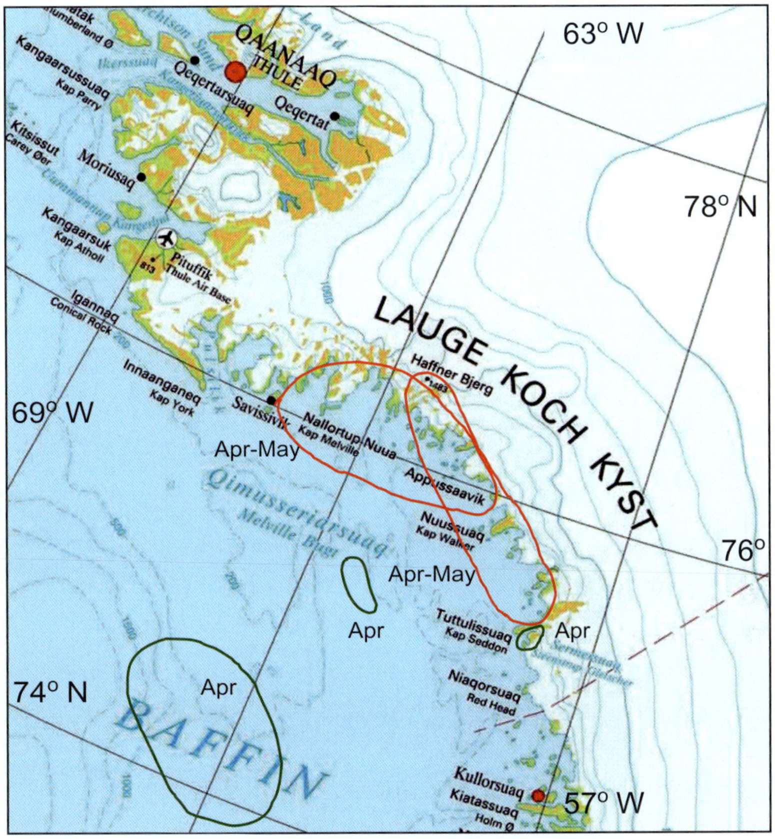

FIG. 21. Areas where bear tracks are often observed in Qimmusseriarsuaq/Melville Bay (Area 3) as drawn by four hunters during the interviews in Kullorsuaq and Nuussuaq in February 2006. Red = both sexes and all age classes; green = tracks made by females with cubs.

There were various statements regarding the track direction and the factors that might determine it. Several of the statements indicated that the tracks may go in any direction. Some of the informants also specified that the direction is determined by the direction of the wind (*i.e.* the scent of the prey or mates), the current patterns, and/or the properties

	Number of responses	Kullorsuaq	Nuussuaq
Drift ice to the west	6	5	1
Seen more in the "local" area	4	2	2
Everywhere	4	1	3
Other places	2	2	-
N_{total} (responses)	16	10	6

Table 47. Types of responses to question 11, from Kullorsuaq and Nuussuaq (Upernavik municipality): "Where have you seen bear tracks?"

of the ice. One hunter from Kullorsuaq was of the opinion that the bears follow the scent of ringed seal pups in April, but another hunter from the same settlement said that they primarily move in a westerly direction in April. According to another informant, this is also the primary direction of movement immediately after the ice has formed in the autumn. A 66-year-old hunter from Nuussuaq told the interviewer that the tracks of large male bears indicate a northerly direction of travel in April – towards females in heat.

Abandoned sod houses on Tuttulissuaq/Cape Seddon, August 2007. Photo: K. L. Laidre

12. Do the bears have regular migration routes?

A total of 16 interviewees were asked this question: 4 (25%) answered "yes", 8 (50%) answered "no" and 4 (25%) had no particular opinion on the matter (Table 48).

Three of the four respondents who answered in the affirmative said that the bears tend to travel along the aforementioned leads in the ice along the coastline and around areas of shallow water. The fourth informant did not elaborate on his response.

13. Have you seen very small bear tracks (i.e. from cubs of the year)?

Of the 17 respondents, 8 (47%) had seen tracks made by very young cubs and 9 (53%) had not (Table 49).

	Total	Kullorsuaq	Nuussuaq
Yes	4	2	2
No	8	5	3
No opinion	4	3	1
N_{total} (hunters asked)	16	10	6

Table 48. Distribution of the responses to question 12, from Kullorsuaq and Nuussuaq (Upernavik municipality): "Do the bears have regular migration routes?"

Six informants from Kullorsuaq stated that they had seen very small bear tracks. A hunter from this settlement had seen such tracks ("a little larger than dog tracks") out on the drift ice a long way from land. Another hunter from the settlement had seen similar tracks just east of Tuttulissuaq/Kap Seddon in March (1990?). In that instance, an old male bear had attempted to attack a female with cubs but the tracks indicated that the female had protected them. The cubs had fled while the male and female bears fought. Subsequently, the female had caught up with her young but the male followed them and the struggle was repeated. Finally, the female and her cubs managed to escape by fleeing into the mountains.

A 44-year-old hunter from Kullorsuaq spoke of how some people began to suspect that the polar bears had maternity dens around the large icebergs out in the drift ice in Baffin Bay:

> It was also then that ... at the beginning of June ... back when the ice used to be thicker. Back then, we started to suspect that they sometimes had maternity dens by the big icebergs [........] That was back then – maybe in the early 1990s or the late 1980s [........] I thought that because the tracks were too far west, and because they were so small. Because they were so far west that we had driven to the west for three days, from Kullorsuaq. They were smaller than dog tracks.

	Total	Kullorsuaq	Nuussuaq
Yes	8	6	2
No	9	5	4
N_{total} (hunters asked)	17	11	6

Table 49. Distribution of the responses to question 13, from Kullorsuaq and Nuussuaq (Upernavik municipality): "Have you seen very small bear tracks?"

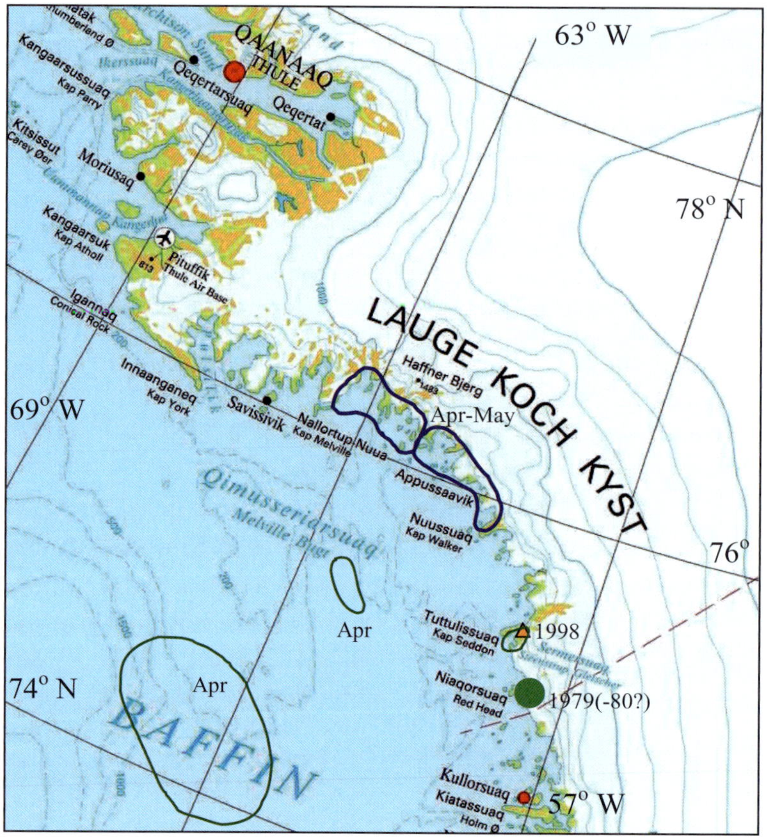

FIG. 22. Observations of tracks made by females with newborn cubs and dens in Qimmusseriarsuaq / MelvilleBay (Area 3) as indicated on a map by two hunters during the interviews in Kullorsuaq and Nuussuaq in February 2006. Blue = general breeding areas on the coast; green lines = tracks made by females with small cubs; orange triangle = temporary dens; green dot = maternity den.

The same hunter pointed out on a map a large area in the northern part of Qimusseriarsuaq / Melville Bay where bears breed (Fig. 22). Another hunter from the same settlement drew areas in Qimusseriarsuaq on the map where he had observed tracks of female bears with very young cubs (*anerlaat* = bear cubs that have just left the den) in April–May, going in a westerly direction (Fig. 21). Without going into details, he mentioned that he had seen maternity dens there and in the Kullorsuaq area. In April 1997 another hunter had seen very small tracks in the Tuttulissuaq area and another hunter reported having seen the same type of tracks by the shore at Niaqorsuaq – an area where dens are said to occur. He had furthermore seen *"sarliarsuit"* (*i.e.* a female bear with two nursing cubs) between *"maniitsorsuit"* (jagged icebergs) in the same area in the late 1980s. According to this hunter, the female with cubs leave their dens at the end of April or the beginning of May.

A hunter from Nuussuaq spoke of how he had seen small tracks in the drift ice out to the west in March. A 66-year-old resident of the same settlement recounted:

> When I was a child, and when we were allowed to hunt any bears no matter how old they were, all year round – at that time the hunters usually came home with two very small, dead bear cubs [........]. If I remember correctly, it was in March. When I used to go west to hunt bears myself, I saw one year old bears there … less than one year old, but hunting these small *"atertallit"* (*atertat* = bear cubs; author's note), as we call them, was not allowed. [........] I have always encountered them.

The very small cubs were observed on the drift ice in April.

14. Where do the bears spend the summer?

Of the 16 informants to whom this question was posed, half of them had an opinion on the subject and the other half did not (Table 50). There was not much clarity regarding this subject, and there were various perceptions of where the polar bears go in the summer (Table 51).

	Total	Kullorsuaq	Nuussuaq
Opinion on this	8	3	5
No opinion	8	7	1
N_{total} (hunters asked)	16	10	6

Table 50. Distribution of the responses to question 14, from Kullorsuaq and Nuussuaq (Upernavik municipality): "Where do the bears spend the summer?

	Number of responses	Kullorsuaq	Nuussuaq
On the drift ice	3	1	2
On land	2	2	-
Drift ice and on land	1	1	-
Near the glaciers	4	2	2
Other	3	2	1
N_{total} (responses)	13	8	5

Table 51. Types of responses to question 14, from Kullorsuaq and Nuussuaq (Upernavik municipality): "Where do the bears spend the summer?"

Several respondents were of the opinion that the polar bears are to be found on the drift ice in Baffin Bay or as one informant expressed it, "on the open sea where we sometimes meet them". Among the informants who thought that the bears are also to be found on or around the glaciers, one mentioned Qimusseriarsuaq/Melville Bay and another said they stay on the fields of sea ice in front of the calving (*i.e.* productive) glaciers. A single informant was of the opinion that the bears may occur anywhere. According to two other interviewees, it has become more common that hunters encounter bears in the summer, while others claimed that this is something which is seldom heard of.

A sedated female polar bear with cubs of the year in Qimusseriarsuaq/Melville Bay (May 1992) where some bears have maternity dens. Photo: E.W. Born

15. Have you seen dens with females with small cubs (maternity dens)?

Of the 17 respondents, only 2 (from Kullorsuaq and Nuussuaq, respectively) said that they had seen maternity dens. A couple of hunters from Nuussuaq did however speak of the observations made by other hunters of a maternity den. Others pointed out areas which are renowned bear breeding areas (Fig. 22).

A hunter from Kullorsuaq had seen a maternity den in May 1979 (1980?) by the Kjer Glacier near to Tuttulissuaq/Kap Seddon in Qimusseriarsuaq. An informant from Nuussuaq recalled how he had seen a female bear with two cubs of the year by a den in Qimusseriarsuaq while he was on the way to Savissivik at the end of April in the late 1980s (from the statement, it seems as though they followed the edge of the land fast ice along the lead *"Qimusseriarsuup Oqaa"* / *"The Tongue of the Sea"*). Another informant from this settlement said that he had heard of how it is still possible to find maternity dens in the vicinity of Nuussuaq in the area with "high snow" – for example the "western islands" and around the glaciers. A 66-year-old hunter told how he had heard of a maternity den which had been observed over 20–30 years ago close to the town of Upernavik.

The accounts given indicate that it is unusual to observe maternity dens, but some hunters also emphasised that they do not explore the potential denning area in Qimusseriarsuaq / Melville Bay, because it is protected.

16. Have you seen temporary dens?

Just five (29%) of the 17 respondents had seen temporary dens (two from Kullorsuaq and three from Nuussuaq) while the remainder had not. The location of temporary dens in the northernmost part of Upernavik was also indicated on a map by the informants (Fig. 22).

Some of the informants provided details of temporary dens. A hunter from Kullorsuaq had three times in April–May seen such dens located in deep snow drifts. He described them as being less tall than a man and maybe 2–3 metres deep (with just enough space for one bear) and with an aeration duct (*qingaq*, "nose"). There were also signs indicating that the bear had a place where it could urinate. Another hunter from the same settlement told of an abandoned temporary den which he had seen in March 1998 (year?) at Tuttulissuaq / Kap Seddon. He had been into the den which consisted of a 10 m (it is not clear whether the informant actually said "ten"; author's note) entrance tunnel followed by an approximately 1 m high chamber with a *qingaq*. A hunter spoke of how others had illegally hunted a bear in its den:

> I haven't seen them. And ... quite a few days ago, I don't know ... I heard some talk about bears that had been digging in the snow. I don't know how true it is. There were bears on land around Kiatassuaq / Holm Island, two bears. Here. A female and her cub were caught. They had apparently dug down into the snow [........] Not that long ago, in January. I think it was in January (2006; author's note), maybe, or was it the start of February, that a female and a cub were caught? Because of confusing information (probably about the bear hunting regulations; author's note) from the Upernavik Municipal Office, illegal bear hunting does take place. That is very regrettable.

Another hunter from this settlement mentioned that a bear had been shot in its den at Innaarsuit "a few days ago".

A hunter from Nuussuuaq had seen temporary dens on and around the glaciers in Qimusseriarsuaq/Melville Bay in August–September. His account indicated that the polar bears spend long periods of time in between the large segments of ice calved from the glacier fronts. Another hunter from the same settlement thought that the denning bears are animals which, "when they get too fat, dig down into the snow and go into the snow in order to lose weight." He had seen dens in snow drifts around the icebergs out in the pack ice to the west of Nuussuaq. A third informant from Nuussuaq said that dens may be observed anywhere but that they are maybe more frequently found along the coastline.

17. Have there been changes to the occurrence of dens over the years?

This question was put to eight informants, three of which (2 from Kullorsuaq and 1 from Nuussuaq) responded by saying "no" (38%), and the remainder had no opinion on the subject. One individual did however remark that the bears' choice of denning location may vary.

Kullorsuaq in early April 2010. Photo: K. L. Laidre

Climate change

18. Have you observed changes to ice conditions?

Sixteen (94%) of the 17 respondents answered this question in the affirmative, while one hunter aged 36 from Kullorsuaq answered "no". The majority of the responses to this question related to the suggestion that the ice forms later and is thinner and more dangerous to travel over. Some also stated that the leads in the ice now occur in different locations and that they do not freeze over as easily as they used to (Table 52).

	Number of responses	Kullorsuaq	Nuussuaq
Forms later and breaks up earlier	13	9	4
Thinner and more unsafe	9	4	5
Holes in the ice and more open water	2	-	2
Changes to the ice edges	-	-	-
Other	-	-	-
N_{total} (responses)	24	13	11

Table 52. Types of responses to question 18, from Kullorsuaq and Nuussuaq (Upernavik municipality): "Have you observed changes to ice conditions?"

Nine hunters (5 from Kullorsuaq and 4 from Nuussuaq) told the interviewers that the ice used to form at the end of October or in November but that in recent years it has not formed until the end of December. Several informants spoke of how it had previously been possible to drive a dog team over the ice in November, but that nowadays this cannot take place until December, and even then they must exercise caution. A number of the respondents recounted how the ice formed early and subsequently remained stable during their childhood, which in this context refers to the period from the 1950s to sometime in the 1980s. Two hunters from Kullorsuaq were of the opinion that the ice had formed relatively early even into the first half of the 1990s and that the change had occurred after that time. According to one of these informants, the last two years was where the ice has formed late (*i.e.* 2003 and 2004; author's note). An informant from Nuussuaq stated that the ice broke up again in 2004 because of rough seas. A hunter from Kullorsuaq mentioned that in 1996 there had been ice coverage as early as October.

A 50-year-old hunter from Kullorsuaq said that his father had told him how ice conditions had also been problematic and uncertain in the late 1920s:

> The day before Christmas Eve. One time there was the sound of the sea. The sound of the sea ... that means that there was no ice. When they went for supplies in Nuussuaq they sometimes returned on Epiphany (6 January; author's note). To the east of here. Then they came back the day after Epiphany or later ... they had just gone away to stock up. Then it started to happen that in the early 1980s we started to have lots and lots of ice. The way I understand it, this happened before climate change. It started with frost and the ice forming so early. Then in the 1980s and the early 1990s, the ice started to form much later.

The instability of the ice makes it difficult to travel over. Two hunters from Nuussuaq stated that in recent years, it had not been possible to drive to the west (*i.e.* out onto the drift ice in Baffin Bay; author's note) because of unsafe ice conditions. According to one of them (age 66) this is in part because the ice may break up and because ice does not form over the leads. In addition to this, he explained that there are now more leads than there were before. According to this man, these changes are the result of climate change. He remarked that the sea has become markedly warmer and that the weather in the autumn is now more erratic. The second informant remarked that in the area around the shore lead at Uummannaq (*i.e.* a mountain at the Nuussuaq peninsula) there had previously been a clear edge, but that in recent years there had been thin, unsafe ice, something which has also been observed in other locations in that area.

A 46-year-old from Kullorsuaq hypothesised over the relationship between the ice, ringed seals and bears:

> There was very thick ice in the area close to us, when I was a child. Very, very thick and stable ice. But now we get very thin ice, which is not as thick as it used to be. Therefore, I think that bears prefer to hunt on thin ice. I think that it is so, because the seals' breathing holes are much easier to find when there is thinner ice.

Two informants also spoke of how the ice foot (*qaanngut*) is formed later and that it is now narrower, with more leads, which among other things means that sections break off.

A 67-year-old from Nuussuaq described how the ice conditions have changed:

> It used to be possible to drive directly to Kullorsuaq from here in December. In recent years that has been completely impossible. Things have changed a lot compared to when we were young, when we could drive directly to Kullorsuaq in December. Now we can't do that until January, we are able to drive to Kullorsuaq from around the end of January nowadays, and even then not by the direct route.

We were also told that the ice has broken up earlier in the spring, in recent years. A 44-year-old hunter from Kullorsuaq thought that changes to the current patterns were of significance:

> ... As a result of the changing current patterns, there is now open water much earlier. I can name for example that back in 1983, when I had just moved here, we didn't use to go hunting on the ice edge until the end of June. We hunt narwhals, for example, from the edge of the ice, but now the ice-edge hunting and other forms of hunting mostly take place at the end of May.

Additionally, a couple of hunters said that in the winter of 2005/2006 had been more normal, with relatively early ice formation and more stable ice coverage.

A 36-year-old resident of Kullorsuaq, who had shot his first bear at the age of 22, did not think that there had been any change to ice conditions since he had started hunting. According to this man, the ice usually forms in December.

19. Have you seen changes to the icebergs?

Nine (ca. 53%) of the 17 respondents answered "yes" to this question and 8 (47%) answered "no" (Table 53).

	Total	Kullorsuaq	Nuussuaq
Yes	9	5	4
No/same	8	6	2
N_{total} (hunters asked)	17	11	6

Table 53. Distribution of the responses to question 19, from Kullorsuaq and Nuussuaq (Upernavik municipality): "Have you seen changes to the icebergs?"

All of the informants who had noticed changes said that the number of icebergs had decreased. A hunter from Kullorsuaq recalled how the glaciers had produced large jagged icebergs (*"maniitsut"*) in the mid 1990s but said that this was no longer the case. Another informant remarked that the icebergs now contain more soil and clay (*"marraq"*) than they did previously.

The absence of stranded icebergs means that there is less protection against the sea. A 66-year-old hunter told the interviewer that:

> The great sea has got warmer. In that way, it melts the large icebergs, so that those that run aground even break up. In the summer of 2003, we could see Uummannaq (the mountain by Nuussuaq; author's note) to the west from the mountain, you know. You have seen Uummannaq. At Ummaannaq there are usually big icebergs. Then, to the east … all of the big icebergs … they have all disappeared. When we were children, there were some islands to the north of Kullorsuaq where there were enormous icebergs. To the west of that which we call Naajatalik. That is why there were no waves in Kullorsuaq back then. They shielded this whole area.

An informant from Nuussuaq spoke of how the icebergs which come from the three glaciers to the east of the settlement are now smaller than they used to be.

A hunter from Kullorsuaq pointed out an area to the northwest of the settlement, where icebergs tend to run aground (Fig. 23).

20. Have you seen any changes to the glaciers?

Of the 17 hunters from Kullorsuaq and Nuussuaq who responded to this question, 16 (94%) answered in the affirmative, while one hunter from Nuussuaq gave a somewhat unclear answer.

A recurrent response to this question (89%) was that the glaciers had clearly receded (*i.e.* towards the east) and some of the answers given indicated that the retreating glaciers have exposed areas of land and *nunataks* (*i.e.* sections of land surrounded by glacial ice). A couple of responses indicated that certain glaciers have increased in productivity (Table 54). Some of the hunters pointed out on a map how the glacier fronts have changed (Fig. 23).

The front of the Steenstrup Glacier, which is situated to the south of Tuttulissuaq/Kap Seddon, has moved further to the west (*i.e.* out to sea) according to a hunter from Nuussuaq. According to two hunters, the edge of the Kjer Glacier has moved in an easterly direction. This means that an area of land has now appeared

FIG. 23
Observations of changes to the glaciers in Qimmusseriarsuaq/Melville Bay (Area 3), as indicated on a map by two hunters during the interviews in Kullorsuaq and Nuussuaq in February 2006. An area with many icebergs (black line) as drawn by a third hunter is also shown.

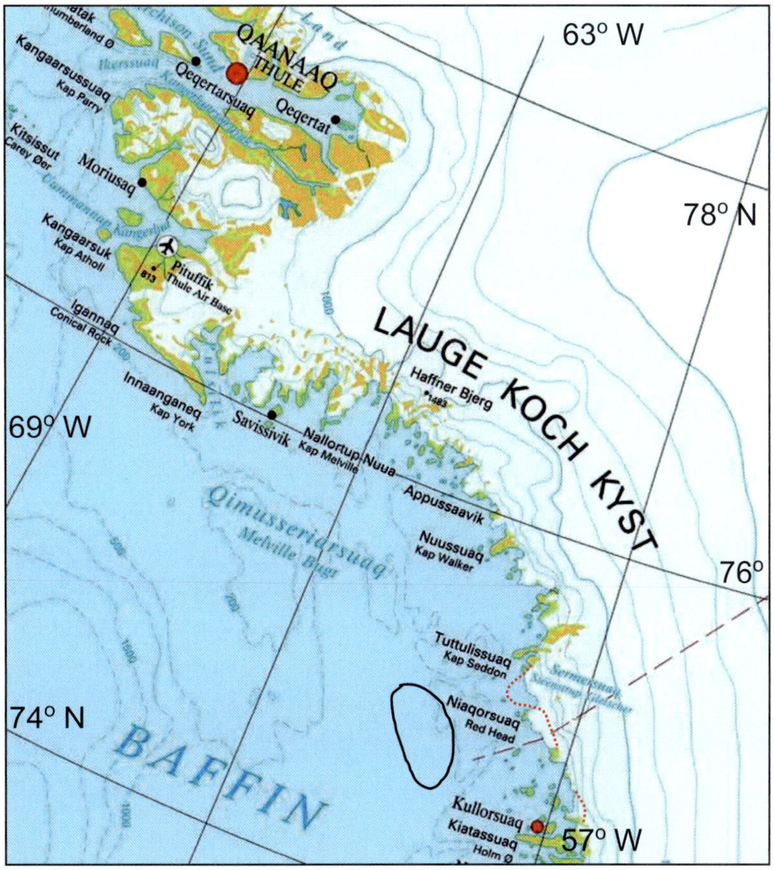

	Number of responses	Kullorsuaq	Nuussuaq
Retreated	16	11	5
More productive	2	2	-
N_{total} (responses)	18	13	5

Table 54. Types of responses to question 20, from Kullorsuaq and Nuussuaq (Upernavik municipality): "Have you seen any changes to the glaciers?"

east of Red Head and Niaqorsuup Saarlia and behind Hovgaard Kystland. The edge of the glacier called Nunatakasaap Sermia, which lies just to the east of Kullorsuaq, has also receded in an easterly direction. A 34-year-old hunter from Kullorsuaq described the change thus:

> For example, close to our fishing spot ... here ... the glacier here. I have more than once gone up to the summit of Naajarsuit. The first time that I crawled up there, it was possible to see the glacier. In August (2005; author's note), I crawled up there, but I couldn't even see the glacier, it had moved so far to the east.

A 47-year-old hunter from Kullorsuaq stated that the changes had taken place during the 1990s. A 50-year-old from the same settlement described how the first signs of change began to take place in the 1980s, as follows:

> The glacier over there, there. Here on the map, here was the glacier (*i.e.* between Wendell Land ca. 74° 35′ N, and Lille Renland ca. 74° 55′ N; author's note). Then, in the early 1980s ... my father shouted out loud, "How come?"... the fact that the glacier moves to the east. You know. There was no response. After that, it became obvious [........] We had actually made enquiries by phone, you know, as to why the glaciers moved east so quickly, you know. What actually causes it? It is only in recent years, you know ... about climate change ... you know ... they open their eyes. We, the hunters, have noticed it since then, naturally because hunters live close to it, you know.

The same man thought that the glaciers calve more violently in the spring now, as well as in the winter when the current is strong. A 36-year-old man from the same settlement said that the glaciers are constantly calving while they move east.

21. Have you seen any changes to the snow?

Thirteen (ca. 76%) of the 17 interviewees answered the question by saying "yes", 3 (ca. 18%) answered "no" or that there had been no change, whereas 1 (ca. 6%) had

	Total	Kullorsuaq	Nuussuaq
Yes	13	8	5
No/Same	3	2	1
No opinion	1	1	-
N_{total} (hunters asked)	17	11	6

Table 55. Distribution of the responses to question 21, from Kullorsuaq and Nuussuaq (Upernavik municipality): "Have you seen any changes to the snow?"

no opinion on the matter (Table 55).

The majority of the responses indicated that less snow has fallen in recent years. Some informants also remarked that the snow falls at different times of year than it used to and that precipitation can come "unexpectedly". One informant did however state that lots of snow had fallen in recent years, but added that in the winter of 2005/2006 there had not yet been much snowfall (Table 56).

	Number of responses	Kullorsuaq	Nuussuaq
Less falls	7	6	1
More falls	1	-	1
Change in the time of year	4	2	2
Also rain	4	3	1
Other	1	-	1
N_{total} (responses)	17	11	6

Table 56. Types of responses to question 21, from Kullorsuaq and Nuussuaq (Upernavik municipality): "Have you seen any changes to the snow?"

A hunter from Nuussuaq was of the opinion that the snow has started to fall relatively late, but according to two hunters from Kullorsuaq, it is normal for the snow to fall in April. Another informant from Nuussuaq had noticed that nowadays large amounts of snow can fall in a short period of time whereas the snow used to fall over a longer period.

Three hunters from Kullorsuaq explained that in recent years, they have experienced rainfall at unusual times. One of these informants said that it can even rain in the middle of winter, while another man from the same settlement had experienced rain falling just after snow. The same phenomenon of snow followed by rain was also mentioned by a hunter from Nuussuaq (see also the answers to question 23, p. 117).

A statement made by a 50-year-old hunter from Kullorsuaq did however indicate that a certain amount of variation in the precipitation during the winter period had also occurred in the past:

> During our childhood (in the 1960s; author's note), it rained then. During the month of February, sometimes. The weather usually got milder in February, you know. Then we usually had snow in April. At around the end of April [........] It tends to snow in the autumn.

> From October, about when the dark period begins, there is usually a lot of snow. But in recent years … in the last two years we haven't had snow. We can just say that … now … it is strange, isn't it?

A 66-year-old hunter from Nuussuaq was of the opinion that the amount of snow varies cyclically:

> The snowfall is changeable. Maybe 13–14 years at a time. Sometimes, the snow is much thicker. And in other years, there is very little snowfall.

22. Have you seen any changes to current patterns?

Eleven (65%) of the 17 respondents answered the question by saying "yes", 2 (12%) said "no" and 4 (ca. 23%) had no opinion on the subject (Table 57).

Some of the responses that were accompanied by an explanation indicated that the current had increased in strength (Table 58).

Four informants (3 from Kullorsuaq and 1 from Nuussuaq) were of the opinion that the currents had changed because the glaciers have receded and thereby creating new straits. Seven hunters (3 from Kullorsuaq and 4 from Nuussuaq) spoke of how changes to the current have meant that the ice no longer forms as thickly as it used to and that as a result of this, dangerous areas of poor ice as well as holes

The settlement Nuussuaq (formerly called Kraulshavn) on the Nuussuaq peninsula in the northern Upernavik area, April 2010. Photo: K. L. Laidre

	Total	Kullorsuaq	Nuussuaq
Yes	11	6	5
No/Same	2	2	-
No opinion	4	3	1
N_{total} (hunters asked)	17	11	6

Table 57. Distribution of the responses to question 22, from Kullorsuaq and Nuussuaq (Upernavik municipality): "Have you seen any changes to current patterns?"

	Number of responses	Kullorsuaq	Nuussuaq
Have changed	7	4	3
Increased in strength	4	2	2
N_{total} (responses)	11	6	5

Table 58. Types of responses to question 22, from Kullorsuaq and Nuussuaq (Upernavik municipality): "Have you seen any changes to current patterns?"

(*aakkarnerit*) are formed. An individual from Nuussuaq also remarked that the sea had got warmer.

A 40-year-old from Kullorsuaq, who had answered the question by saying "no" had only observed the usual changes to the strength of the current, which are due to the lunar cycle.

23. Have you seen any changes to the weather?

Of the 17 interviewees who were asked this question, 14 (ca. 82%) answered "yes", 1 (6%) said "no" and 2 (12%) had no particular opinion on the matter (Table 59).

	Total	Kullorsuaq	Nuussuaq
Yes	14	10	4
No/Same	1	-	1
No opinion	2	1	1
N_{total} (hunters asked)	17	11	6

Table 59. Distribution of the responses to question 23, from Kullorsuaq and Nuussuaq (Upernavik municipality): "Have you seen any changes to the weather?"

	Number of responses	Kullorsuaq	Nuussuaq
Warmer	9	6	3
More wind	1	1	-
Unstable / rainy	6	4	2
Other	1	1	-
N_{total} (responses)	17	12	5

Table 60. Types of responses to question 23, from Kullorsuaq and Nuussuaq (Upernavik municipality): "Have you seen any changes to the weather?"

As in the Qaanaaq municipality, the informants from this part of the Upernavik municipality had noticed that the weather has been warmer in recent years and that it has become less stable (Table 60).

An informant from Kullorsuaq stated that during the 1990s, the weather had become warmer with more rainfall. In Nuussuaq, two hunters (aged 66 and 67, respectively) said that the temperature had been under ca. -40 °C on occasion in the 1960s, and one of them added that the temperature is now milder around -20 °C in February. A 63-year-old hunter from Kullorsuaq stated that the weather has become stormier. Furthermore, a number of informants said that the weather has become less stable with sudden changes in temperature and rain during the winter (Table 60). A resident of Kullorsuaq told the interviewer that it had rained a great deal in January–February 2005, and a hunter from Nuussuaq recounted how in recent years during the autumn the weather has alternated between warm and cold until the cold set in, in December.

A 50-year-old from Kullorsuaq mentioned changes to the direction of the wind:

> Regarding the time when it freezes and regarding the wind directions, I have noticed something. In the past, it was the case that the wind usually came from the west with a great force, you know. Sometimes for days at a time. Sometimes we had a strong westerly wind over a period of several days. Nowadays, there is only a light westerly wind [........] And then the north wind. I have also noticed that [........] it is roughly like a westerly wind, isn't it. A north-westerly wind, you know. Now there is very frequently strong wind. This settlement has now got further west, in relation to the glacier. Particularly from the end of June until August, you know.

24. Have you observed any other changes?

Five (29%) of the 17 respondents answered "yes", 7 (ca. 42%) answered "no" and 5 (29%) had no opinion on this matter (Table 61). Four of the informants who had observed changes talked about seals and one talked about narwhals (Table 62).

A 40-year-old informant from Kullorsuaq had remarked how the hunters caught a large number of seals before it became possible to trade Greenland halibut to the local freezing plant (*i.e.* early 1990s). However, ever since fishing activities have intensified there have seemingly been a greater number of ringed seals.

A 66-year-old hunter from Nuussuaq mentioned hairless seals:

> [........] now, when we sail in boats in the west all year round, we have begun, during our travels these days in these years, to see seals with very poor coats. We call them *mamaartut*. They shed their coats. These can be seen all year round, though...[........] And sometimes

	Total	Kullorsuaq	Nuussuaq
Yes	5	2	3
No	7	5	2
No opinion	5	4	1
N_{total} (hunters asked)	17	11	6

Table 61. Distribution of the responses to question 24, from Kullorsuaq and Nuussuaq (Upernavik municipality): "Have you observed any other changes?"

	Number of responses	Kullorsuaq	Nuussuaq
Bears	-	-	-
Other seals and walruses	4	1	3
Whales	1	1	-
Birds	-	-	-
Other	-	-	-
N_{total} (responses)	5	2	3

Table 62. Types of responses to question 24, from Kullorsuaq and Nuussuaq (Upernavik municipality): "Have you observed any other changes?"

> they have almost no fur. *Allattuuaqqat*, young harp seals. In the south they call them *aataannguit* [........]. In the past one never saw them. You know. Now we catch some which are almost hairless. But last autumn there were seals that were moulting, and some of them were sent away to be examined [........].

A 67-year-old informant from the same settlement mentioned that in 2005 seals had not been seen for most of the summer, whereas after the frost had come "earlier than expected" there had suddenly been a large number of seals. A 66-year-old hunter from Nuussuaq wondered whether the polar bears eat the small ringed seals (young seals?; author's note):

> Then, in 1997 we set nets out there [........], even as early as February there were very few of these small ringed seals left. I think that the reason is that the population of bears was so small here, in the 1960s. Large numbers of these ringed seals were born. In the autumn, in October, to the north of Nuussuaq, then they come towards the shore here in October. Ringed seals en masse. Now there are none of them left. The bears here in Baffin Bay, which number more than 2000, have eaten them all. This year, you can ask some people about the size of the ringed seals that are caught. Now only the big ringed seals are left. It's unbelievable.

Without going into details, a 44-year-old informant from Kullorsuaq voiced the opinion that narwhals have not, "as some people have claimed", decreased in number, but that they have in fact just changed their migration routes. This man also mentioned that there are now more common eiders (*Somateria mollissima*) in the Kullorsuaq area and that it has been said that there are generally fewer eiders in the south and more in the north.

Changes to the catch and occurrence of polar bears

25. Have the changes affected the polar bear hunt?

This question referred specifically to the extent to which changes to the physical environment that are related to the climate (ice, wind, glaciers, weather etc.) have affected the bear hunt.

Of the 16 informants who were asked this question, 9 (56%) answered "yes", 5 (31%) answered "no" and 2 (13%) had no opinion on the subject (Table 63). A

	Total	Kullorsuaq	Nuussuaq
Yes	9	5	4
No / Same	5	4	1
No opinion	2	1	1
N_{total} (hunters asked)	16	10	6

Table 63. Distribution of the responses to question 25, from Kullorsuaq and Nuussuaq (Upernavik municipality): "Have the changes affected the polar bear hunt?"

	Number of responses	Kullorsuaq	Nuussuaq
Closer	1	1	-
- due to changes in the ice	2	2	-
Changed methods	6	2	4
Other	-	-	-
N_{total} (responses)	9	5	4

Table 64. Types of responses to question 25, from Kullorsuaq and Nuussuaq (Upernavik municipality): "Have the changes affected the polar bear hunt?"

third of the affirmative answers indicated that the bears had come closer to coastal areas, while the remainder contained information on how the changes had affected hunting patterns (Table 64).

Some informants said that the bears occur closer to the coast. A 40-year-old informant from Kullorsuaq was of the opinion that the bears generally occur in the coldest areas. According to him, they rest when the temperature is too high and move around more in cold temperatures. This is also why they mostly move around at night in the spring. He was therefore of the opinion that the bears, in reaction to the "changing weather", are naturally to be found on occasion closer to land where it is cooler in the proximity of the glaciers.

A 34-year-old from the same settlement pointed out an association between the distribution of ice, travelling conditions and the occurrence of polar bears. He said that in the past when there had been a lot of ice, it had been necessary to drive a dog team a long way in order to encounter polar bears but that nowadays they can be encountered after having travelled just a short distance.

A 46-year-old hunter from Kullorsuaq philosophised over the extent to which there had been an increase in the bear population or a change in their distribution:

> I believe that they (the changes; author's note) improve the bear hunt, you know. That the ice comes later, and the bears come closer. I don't know how this is to be understood. Whether they are increasing in number or just coming closer to land. I cannot understand this. I really don't know. I cannot even guess whether there are more bears now, or whether they are just coming closer to the shore.

However, most of the respondents to this question spoke of how the changes had had an influence on hunting patterns. In particular how the lack of sea ice or reduced sea ice impedes the use of dogsleds for hunting yet results in an increase in skiff hunting. A 44-year-old hunter from Kullorsuaq said:

> Here it is of course the ice …whether the ice comes earlier and whether there is open sea earlier … that can have an effect on bear hunting from dogsleds. Maybe because we get open water earlier, when we were otherwise able to hunt bears from dogsleds, we travel by skiff, so now hunting takes place from skiffs.

Without elaborating on his statement, a 63-year-old from the same settlement mentioned that the late formation of the sea ice adversely affects the bear hunt. Four hunters from Nuussuaq likewise explained how the unsafe ice conditions have affected the hunt in various ways, not least the bear hunt on the drift ice "out west" which was mentioned by several respondents. A 42-year-old responded to this question as follows:

> Not them exactly. I don't think that they affect anything (*i.e.* the climatic changes; author's note). It is only the western ice. We travel west exclusively to hunt bears. It affects it [........] These days we cannot go west exclusively to hunt bears because we have no routes to travel on [........] It did not use to be like that in the past. There used to be ice.

A 66-year-old from the same settlement expressed a similar opinion, as follows:

> The areas where they used to go to hunt polar bears, they are inaccessible nowadays. Lots of bears are caught. They migrate all the time, you know. You can come across them anywhere.

A 67-year-old informant from Nuussuaq described how it is generally the case that the bear hunt begins earlier in the years when the ice forms early, while the opposite is true in the years when the ice forms late.

26. Have there been any changes to the polar bears that you have seen or caught?

The intention with this question was to glean information regarding the extent to which bears had become thinner, fatter, or changed in physical condition.

Four (25%) of the 16 informants who were asked this question answered "yes", 11 (69%) answered "no" and 1 (6%) had no particular opinion on the subject (Table 65).

	Total	Kullorsuaq	Nuussuaq
Yes	4	2	2
No/Same	11	7	4
No opinion	1	1	-
N_{total} (hunters asked)	16	10	6

Table 65. Distribution of the responses to question 26, from Kullorsuaq and Nuussuaq (Upernavik municipality): "Have there been any changes to the polar bears that you have seen or caught?"

A 34-year-old hunter from Kullorsuaq had noticed a change in colour:

> For example it is as if they are brown this winter. Maybe because the snow ... I don't know. Their fur is browner. As if ... they are not as white [........]. I have caught a bear this winter. As though ... it was a bit brown ... it was really quite brown. In the past they were very white. This might be because there is not so much snow on the ice. I have only noticed it this winter. As though ... the one we caught was so brown.

A 36-year-old man from the same settlement was of the opinion that the polar bears have got "rounder ... they are so fat, maybe because their food is so good."

The two hunters in Nuussuaq who thought that the bears have changed were of the opinion that they have become thinner. One of these informants was a 42-year-old man who is considered by everyone in the settlement to be the hunter who has shot the greatest number of bears. He said:

> In recent years, most of them have been thin [........]. Yes. The bear cannot feed out at sea. So it comes onto the ice to feed, you know. Because there is now open sea, the places where it otherwise used to stay, there where it used to feed … now these areas do not easily form ice. Most bears are thin. They are getting thin. So when they cross over to here from over there (from the drift ice to the west; author's note), then they start to come here between the land [........] They lose their layer of blubber [ ] Yes well, some of the ones that I have caught, they have been like that. They have started to be like that in recent years [........] They weren't like that in the past. They used to be very fat in the past. Of course it is not each and every one that is like that. But for most of them, their layer of blubber has got thinner. Some of them have almost no blubber. This winter I caught yet another one without very much blubber.

This view was supported by a 66-year-old hunter from the same settlement:

> In recent years, some of them have got thinner. There is evidence of this [........] We examine the contents of the stomach, you know. When the bear is not hungry, where there are plenty of seals, then it eats only blubber. When it catches ringed seals, it never eats the meat, only the blubber. Now the bears that are caught sometimes have nothing in their stomach. And they are thin.

Polar bears can swim under the sea ice and break breathing holes with their head in ice that is at least 5 cm thick. Photo: K. L. Laidre

Among the informants who answered this question in the negative was a 46-year-old from Kullorsuaq. He is thought to have caught a total of around 200 polar bears and had never caught a thin bear. A 66-year-old man from Nuussuaq said that some bears are thin and others are fat.

27. Have you observed changes to the occurrence of polar bears?

Of the 16 respondents to this question, 13 (ca. 81%) said "yes", 1 (6%) said "no" and two others had no opinion on this subject (Table 66).

Several of the responses were statements which indicated that the bears have come closer to the coast, but several of them implied that there are more bears now. Some of the responses were a combination of these two statements (Table 67).

Among the informants who said that the bears occur closer to land, a 66-year-old from Nuussuaq, stated:

> [........] in the 1960s there were far fewer bears than there are now. Since the start of the 1980s, I have noticed that lots of bears have

	Total	Kullorsuaq	Nuussuaq
Yes	13	7	6
No/Same	1	1	-
No opinion	2	2	-
N_{total} (hunters asked)	16	10	6

Table 66. Distribution of the responses to question 27, from Kullorsuaq and Nuussuaq (Upernavik municipality): "Have you observed changes to the occurrence of polar bears?"

	Number of responses	Kullorsuaq	Nuussuaq
Closer	4	2	2
- due to changes in the ice	1	-	1
- because there are more of them	9	6	3
Other	2	-	2
N_{total} (responses)	16	8	8

Table 67. Types of responses to question 27, from Kullorsuaq and Nuussuaq (Upernavik municipality): "Have you observed changes to the occurrence of polar bears?"

> been seen close to us. In the north, from Upernavik town up, there are lots of bears.

A 42-year-old hunter from the same settlement expressed the opinion that the decline in sea ice is responsible for a change in distribution, which makes it seem as though there are now more bears:

> Yes. The number of bears is increasing, isn't it? If the western ice (*i.e.* the drift ice in Baffin Bay; author's note) had always been there, then the bears that are heading east would always have been on the western ice [........] Then we wouldn't have seen them [........] We cannot say that they have increased in number. The ones over there (bears; author's note), you know ... that are there [........] Over there, where the ice is breaking up, because they can no longer remain over there, they have therefore started to head east. Therefore it has been said of them that they are increasing in number. They are not increasing in number. They just come here to look for the ice so they can find food. The ones over there have always been there. Those that we never used to see. So there are more now. That's what we say about them. Because they have no place to stay over there. So, in order to look for food on the ice [........] They are increasing in number [........] Yes. The ones that come. From over there.

However, several of the informants answered the question regarding the change in occurrence by saying that the number of polar bears has increased. A 50-year-old from Kullorsuaq was of the opinion that the increase is due to the fact that bear cubs less than one year old became protected (since 1 January 1975; author's note). A 44-year-old hunter from the same settlement explained the increase by saying that it is a response to Canadian hunting:

> If I compare it to how things were in the past, I have really noticed that there are more bears now, I must say. There must be more, because the catch has increased in recent years. And if we say now that the bears may be caught from snowmobiles and other vehicles in Canada, then the consequence may be that they leave this area and head for Greenland. As I said earlier, we maybe had less bears because the ships sailed back and forwards (probably during the construction of the Thule Air Base in the 1950s; author's note) but when the ships disappeared, the bears did return.

A 47-year-old informant, who was also from Kullorsuaq and thought that the number of bears had increased, spoke at length in his response about the effect of an increase in the number of hunters and the use of boats in bear hunting:

> [........] Here I trust what the hunters say, the other hunters. That they are increasing in number. They see a lot of tracks. They have also noticed that there are more now. It is unchanged. It is the same [........] That is quite clear! Otherwise, I really don't know ... or whether they have all moved towards the coast at the same time [........]. Therefore one can maybe say that there is no problem whatsoever, open up the hunt provisionally, it might be easier. But on the other hand we humans are increasing in number. [Laughter]... there are now more who have children. All over the place. There are also more hunters. Maybe it is better that the hunters stick to the hunting licence for the time being. But of course without completely excluding the recreational hunters. But I always say – in the summer – that hunting from skiffs and cutters, I would cite that as the greatest depleting factor on the bear population. But this kind of hunting like it was in the past, from kayaks and dogsleds. It is better not to stop that.

A 37-year-old hunter from Nuussuaq was of the opinion that the bears arrive later because the ice has become "thin". In the past, they used to come in December according to this informant. A 67-year-old hunter expressed his understanding of the situation rather poetically, as follows:

> Yes, there are changes. Because, as I have already said ... the bears were not often seen here in the past. But now that they are so visible they are a source of wonder.

However a 34-year-old informant from Kullorsuaq did not think that there had been any change to the occurrence of bears:

> [........] I have not noticed that there have been any changes to the occurrence of bears [........] In my experience it has remained unchanged. Every year, it is as though ... the bears move about in the area close to us, around on the new ice [........] For example, when the ice starts to be good to move about on, a long way to the west, you know, in about January. Then we start to catch a lot of bears. When we start to make the most of the returning light.

The biology and behaviour of polar bears

28. Have you seen mating, mating behaviour or any tracks that indicate these behaviours?

None of the 16 who were asked this question had observed actual mating behaviours, but three hunters (two from Kullorsuaq and one from Nuussuaq) had seen signs of mating behaviours or tracks indicating mating. Furthermore, a hunter from Nuusssuaq told the interviewer about another hunter's observation of mating.

A 50-year-old informant from Kullorsuaq had seen bears fighting over a female in mid April:

> I have seen mating. But the bears … how … I have also seen bears fighting over a female. We observed an enormous bear without managing to get close to it. Like … there was a lead here … it was here then … a lead … and then on the other side [........] Three bears. A little female [........] It was truly unbelievable! They were fighting over her! There was a cloud of snow over the ice! [........] Three males. [........] They were fighting over a female. One of them was an enormous bear. They fought against this enormous bear, the two of them. They must have injured each other, without a doubt. They must have injured each other badly, no doubt. I think they injured each other. The bear doesn't have much blood, you know. They must die en masse when they injure each other, that's for sure. I have also seen some handicapped bears. For example. This area here (he makes a gesture indicating the area around the lips; author's note) [........] destroyed, some of them without ears. I have also caught bears, some of which had a crushed eye [........] they had not been born like that, though … if they had been born that way … if they were born like that, then they were maybe … during their fights, you know … if we catch some old bears, and we butcher them, right. Scars, scars, scars, all over them. The area here, right (he indicates the areas on his own body; author's note) [........] Mmmm. The area around the face, you know [........]. Especially the legs. With regards to when they fought each other [........] We just turned our backs on them (the three males; author's note), when they behaved like that. The female just went her own way too, you know. [........] I think that the enormous male stopped at the two males. It was after them [........]. I have never seen anything like it, right. They stood upright and did this. It lasted a long time [........] They fought him, and at times the big one was knocked

> down by the others, one of the others, you know. They knocked it down every now and then, and then it got up ... sometimes... I have never...[........] I think the big one won, they were still fighting it. The big one blocked their path. It was much, much, much larger than the others, that one. It cut them off, those two, they had otherwise attempted to beat the big one. I would imagine.

A 47-year-old hunter from Kullorsuaq told of the rutting male bear's behaviour towards dogs:

> Only from tracks. I have seen those that have mated ... that have followed each other. They had maybe travelled together. A married couple. But I have never seen actual mating. But I have been told the following: Dogs are lost to mating bears. Hehehehe ... dogs are crazy about bears after all, you know. In that way. So the mating bears do the following: the male allows the female to go in front, the large one fights with the dogs. [........] It fights with the dogs. It attacks the dogs [........] The male. It lets the female go in front. That's how clever they are. Then, when it is satisfied with the distance, it jumps over the dogs and runs away. And goes over to the female. That's what it does. [........] Yes, it lets its wife get away and then follows her again [........] in its eagerness to mate with her!

Three of the four observations of mating bears or of signs of this were made in April, and one was reportedly made in March–April.

29. What do the bears eat, apart from ringed seals?

Of the 16 informants who were asked this question, half (50%) were able to specify foods other than ringed seals, while the remainder had no opinion on the matter or confirmed that ringed seal is the bears' main food (Table 68).

	Total	Kullorsuaq	Nuussuaq
Yes	8	5	3
No opinion	8	5	3
N_{total} (hunters asked)	16	10	6

Table 68. Distribution of the responses to question 29, from Kullorsuaq and Nuussuaq (Upernavik municipality): "What do the bears eat, apart from ringed seals?"

Remains of a bearded seal pup found in the offshore pack ice in Northwest Greenland in April 2009. A polar bear had eaten the skin and blubber. Photo: E.W. Born

A number of the responses included information about bears eating grass or seaweed (*equutikut*) (Table 69). One informant described how bears even make a hole in the ice in order to eat seaweed by an iceberg or the beach. Others spoke of having found berries, plants and grass in bears' stomachs – even in bears that were well fed.

Some informants mentioned that polar bears eat bearded seals (*taqammutsit, ussuit*) and according to one of them, this is observed on the drift ice to the west

	Number of responses	Kullorsuaq	Nuussuaq
Grass, plants, moss	4	2	2
Seaweed	3	2	1
Other seals and walruses	2	1	1
Whales	4	3	1
Bear cubs	1	1	-
Other (*e.g.* birds, fish)	3	2	1
N_{total} (responses)	17	11	6

Table 69. Types of responses to question 29, from Kullorsuaq and Nuussuaq (Upernavik municipality): "What do the bears eat, apart from ringed seals?"

in particular. Others spoke of how, in certain situations, the polar bears kill ringed seals and then just leave them on the ice (see also question 30). The fact that bears mainly eat the seals' blubber, even when they kill bearded seals, and only eat the meat when they are especially hungry was also mentioned by informants in this area.

Four responses provided information about narwhals. A hunter from Kullorsuaq said that some of the stomachs of the bears that had been caught only contained the outer layer of the narwhals mattak (*i.e.* the skin) – the layer which is known as *meqqua*. According to him, it was as though the bears had gnawed and removed the outer layer without taking the blubber. Another hunter said that bears sometimes occur around narwhal-*sassat* (*i.e.* whales trapped in a hole in the ice), and another cited young narwhals (*uiat*) as bear food. It was reported, that the bears are able to drag young narwhals up onto the ice in order to devour them.

In this area too, there was talk of infanticide and cannibalism. A hunter from Kullorsuaq had seen tracks indicating that a male polar bear had attacked and eaten cubs. He thought that the males actually "decimate the population of mothers with two cubs". Another informant was able to describe a similar observation.

Among other foods, fish and mussels were mentioned. In the case of the latter, a hunter had seen a place where a bear had vomited mussel shells after having eaten their contents (the species was not specified; author's note).

There was however one informant who was of the opinion that a hungry bear is completely indifferent to what it eats and that as a result its prey can include dogs and humans.

30. Have you seen polar bears hunting?

Of the 16 hunters who were asked this question, 10 (63%) answered "yes", while 6 (37%) answered "no" (Table 70). The responses included information about observations of bears hunting ringed seals at their breathing holes (*kikkuleq*; seal's breathing hole), in birth lairs or stalking *uuttut* – *i.e.* seals on the ice (Table 71).

	Total	Kullorsuaq	Nuussuaq
Yes	10	7	3
No	6	3	3
N_{total} (hunters asked)	16	10	6

Table 70. Distribution of the responses to question 30, from Kullorsuaq and Nuussuaq (Upernavik municipality): "Have you seen polar bears hunting?"

Table 71. Types of responses to question 30, from Kullorsuaq and Nuussuaq (Upernavik municipality): "Have you seen polar bears hunting?"

	Number of responses	Kullorsuaq	Nuussuaq
Ringed seals	5	2	3
- at breathing holes			
- in lairs	1	1	-
- as *uuttoq*	2	2	-
Other	1	1	-
N_{total} (responses)	9	6	3

There were reports of bears hunting seals at their breathing holes. A hunter from Kullorsuaq had from a long distance seen a bear sneaking up on a seal which kept peeking out of its breathing hole every now and then. The bear crept forward in a crouching position and it seemed like it moved forward by dragging itself with its claws. It attacked by jumping and then rolled over without effect. Perhaps it had tripped. In order to indicate how strong these bears are, the hunter told of how he had on a number of occasions seen bears crawl up the face of a completely smooth iceberg using their claws. Another man from the same settlement had two or three times seen polar bears which had caught and dragged up seals that were on their way down into their breathing holes.

A hunter from Nuussuaq described the still hunting at breathing holes as follows:

> The bear lies down right in front of the breathing hole and waits until the seal comes up for air. Then just as the seal is about to breathe and the small breathing hole has small pieces of ice and it makes a little current in the hole, so that a little wave sound can be heard ... when the seal starts to do that, then the bear is ready to slap him with its forepaw. [........] with its left [........] Even before the seal surfaces, I made the bear run away (because he was hunting it; author's note).

A 65-year-old from the same settlement had seen that the breathing holes tend to lie to the left of the impression made by the waiting bear in the snow, and had therefore deduced that the bears use their left paw. According to him, the bear digs a shallow hollow with a hole for its head to the right of the breathing hole. It can then catch the seal from below, through the hole (the explanation is somewhat unclear; author's note).

A hunter from Kullorsuaq described how the bears hunt seal pups in their lairs:

> They smell the ringed seals, you know [........]. Depending on the direction of the wind, they wander towards the wind and sniff out the direction. They hunt according to the direction of the wind. They seek out food, these animals. And then they just do this, *aninerit* (seal pups). When the *aninerit* come in April. [........] They start to look for ringed seal pups in April. Sometimes, we come across small pups, small seal pups which have just been killed ... that the bears have killed. [........] Then we just take them, their little heads have just been crushed by it. [........] It eats some of them, if it is hungry. If it isn't hungry, then it just kills them. [........] There is an iceberg here (he draws; author's note). It has snowed and the snow falls in the lee of the iceberg and the ringed seal is in there. Underneath ... in the darkness, perhaps. When it realises this, then the seal digs out a cavern in the snow, from underneath the ice. It gouges out the snow, and makes a lair. Then it stays there [........] And then the bear drags the pup up with its claws. That's how clever the bear is.

Bears had also been observed hunting *uuttut*. A hunter from Kullorsuaq had once in June in the late 1990s observed how a female bear with large cubs stalked an *uuttoq*. While the female crept up on the ringed seal, she held the cubs back by gently swiping at them to stop them:

> They stood quite still and waited patiently. At first, they tried to follow her. But they too had plenty of seal in their stomachs. They had caught a lot of seals and their bellies were full.

Another hunter from the same settlement spoke of how the polar bear attempts to camouflage itself during the stalking:

> When the bear is about to approach the ringed seal ... for example, the bear is white, isn't it, but it is a bit brownish. [........] Then the bear approaches the seal ... so that the colours in the background are similar to the bear. It has been said that they do that. As if they hunt with an eye for that detail. And also ... small high spots, for example small lumps of ice. They approach the seals, hiding behind these high spots all the while. Then, when they are about five or ten metres away, the bear runs at the seal. For example, when the

> seal lays down its head again after having scanned the area. Then the bear runs and when it is about five metres away, then it jumps up and lands on the ice. They jump on the seals [........] Then, when they have jumped on the seal, they throw it backwards [........] Some of the seals are not even touched and it eats only a little of some of them, and then leaves them.

A hunter from Kullorsuaq had observed a bear running around on the ice after seals at Saattut / Sabine Islands in Qimusseriarsuaq / Melville Bay.

31. Have there been any changes in what the polar bears eat?

The responses to this question made it clear that no changes had been observed. The question was asked to 15 hunters, 10 (67%) of which answered "no", while the remaining 5 had no opinion on the subject (Table 72). The responses contained only statements indicating that the bears eat seals, and one informant mentioned that the bears perhaps occur closer to inhabited areas when they are hungry.

	Total	Kullorsuaq	Nuussuaq
Yes	-	-	-
No	10	7	3
No opinion	5	3	2
N_{total} (hunters asked)	15	10	5

Table 72. Distribution of the responses to question 31, from Kullorsuaq and Nuussuaq (Upernavik municipality): "Have there been any changes in what the polar bears eat?"

Area 4 (The area from the south of Nuussuaq to Upernavik town)

In this area, a total of 29 hunters were interviewed (Table 1). The responses to questions 1 and 2 ("How many bears were, in your opinion, caught in your town/ settlement in 2005?" and "How many bears were, in your opinion, caught in the municipality as a whole?") are summarised in the chapter "Number of animals taken" in "The catch in figures", p. 193.

Hunting and travelling conditions

3. Do you catch more bears than you used to?

Three (20%) of the 15 hunters who were asked this question answered "yes", 10 (67%) answered "no", and 2 (13%) did not express any particular opinion on the matter (Table 73).

	Total	Nutaar-miut	Innaar-suit	Naajaat, Tussaaq Aappilattoq	Upernavik town
Yes	3	-	2	1	-
No/Same	10	2	1	5	2
No opinion	2	-	1	1	-
N_{total} (hunters asked)	15	2	4	7	2

Table 73. Distribution of responses to question 3, from the area to the south of Nuussuaq (Area 4) in the Upernavik municipality: "Do you catch more bears than you used to

Only one of the three hunters, a 40-year-old man from Aappilattoq, said that they now catch more bears than they used to and elaborated on his answer. According to this man, who mostly hunts from skiffs and cutters, there had been an increase in the catch in May–June in particular when the bears are at a distance of less than 30 km from the shore (see also Fig. 37). Among the informants who responded negatively to this question, there were some who said that the catch varies from year to year.

4. Are more bears caught here in the settlement than previously?

Thirteen (72%) of the 18 hunters who were asked this question answered "yes", while five (28%) answered "no" (Table 74).

	Total	Nutaar-miut	Innaar-suit	Naajaat, Tussaaq Aappilattoq	Upernavik town
Yes	13	-	5	7	1
No/Same	5	2	1	-	2
N_{total} (hunters asked)	18	2	6	7	3

Table 74. Distribution of responses to question 4, from the area to the south of Nuussuaq (Area 4) in the Upernavik municipality: "Are more bears caught here in the settlement than previously?"

Several informants gave explanations for their response. A 27-year-old hunter from Innaarsuit stated that bears have become "more visible" in recent years. A 40-year-old man from the same settlement said that the bears had been rare in the 1970s, but that they had increased in number during the 1980s and 1990s, reaching a peak in 2000. However, another resident of Innaarsuit (aged 42) claimed that this increase had happened after 2000. A 73-year-old hunter from that settlement pointed out that the increase in the catch is in part due to an increase in the number of hunters (see also the answers to question 8).

A statement made by a 48-year-old hunter from Upernavik, who had hunted from the settlements of Tasiusaq and Tussaaq, confirmed that the number of polar bear hunters had increased and that bears had seldom been seen in the 1970s:

> We never saw bears, I can just tell you that [........] We moved here in 1976. Back then, not that many bears were caught. In the northern settlements, the polar bear hunters travelled by dogsled on the ice for several days. They caught bears. But I have noticed that from the end of the 1970s, when I began to hunt from a sled, the catches in the Upernavik municipality became more frequent. The northern settlements caught bears after several days' travel by sled. Especially those in Nutaarmiut. There is a great bear hunter there. In that way, I have noticed that the bears are increasing in number. Then in later years, during the 1980s ... then they got even more frequent towards the 1990s. From the 1990s to 2000 they became so frequent that they now come directly into the settlements. We have also started hunting the bears like other animals, from skiffs. Maybe even without it being their main occupation. That is, also those who do not have hunting as their main occupation have begun to join in the hunt.

A 46-year-old hunter from Aappilattoq was of the opinion that the increased catch is due to the fact that hunters have begun to look actively for bears:

> Previously, they were not sought out. So they were here all the time, these bears, without us knowing about it. It is only since they began to learn where they are that they have caught them. That is not that extraordinary.

Among the informants who did not think that there had been a change to the catch, there were also some who expressed that the numbers caught depend on the level of the hunting effort. A 66-year-old hunter from Innaarsuit said that the number of bears taken depends on how many people go bear hunting. This opinion was shared by a 45-year-old hunter from Nutaarmiut. An informant (aged 42) from Upernavik explained that the catch had been especially large in 2003 due to the small amount of sea ice in that year (see also question 8).

5. Do you use regular routes when hunting polar bears?

Sixteen (67%) of the 24 respondents answered "yes" to this question, seven (29%) answered "no" and one (4%) said that he had no opinion on the matter (Table 75).

Both boat and dogsled routes were mentioned and in a number of cases were located in the same place. A hunter from Nutaarmiut uses a dogsled trail that runs in a northerly direction from the settlement via Horse Head Island to the west of Tuttoroortooq Island and continues in a westerly direction to an area of shallow water, where leads in the ice are often found. The location of the trail depends on how safe the ice is. He travels to the same area when hunting from a boat. Another informant from the same settlement described how he first checks his seal nets at Upernaviarsuk Island in the west, and from there travels in a southerly direction out onto the drift ice (Fig. 24).

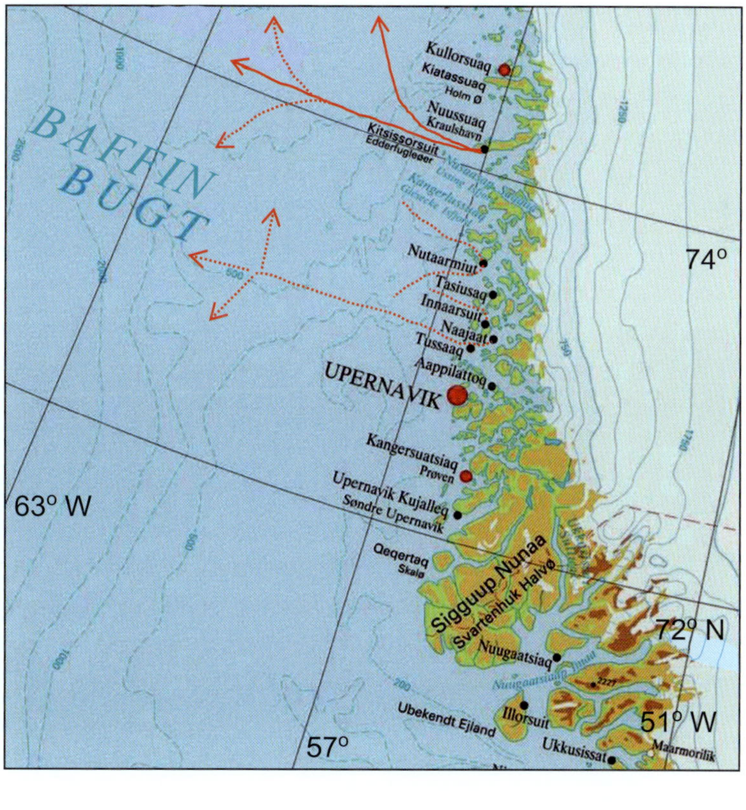

FIG. 24. Sled (solid line) and boat (dotted line) routes in Qimmusseriarsuaq/ Melville Bay (Area 3), as drawn by five hunters during the interviews in Nuussuaq and the southern part of the Upernavik municipality in February 2006. The same routes are often followed whether the mode of transport is a boat or a sled. See also Fig.19.

	Total	Nutaar-miut	Innaar-suit	Naajaat, Tussaaq Aappilattoq	Upernavik town
Yes	16	2	6	5	3
No	7	-	1	4	2
No opinion	1	-	-	1	-
N_{total} (hunters asked)	24	2	7	10	5

Table 75. Distribution of responses to question 5, from the area to the south of Nuussuaq (Area 4) in the Upernavik municipality: "Do you use regular routes when hunting polar bears?"

The hunters from Innaarsuit also usually travel directly to the west and the north-west, and the routes that are chosen depend on where the ice is safe at the time of travel. They head for areas of shallow water with leads and new ice (one informant said that they usually do this in January) – areas that attract the bears. One informant said that on occasion the routes run via the small westerly islands Kingittortaliit (Kingigtuarssuk), and added that in the past, hunters had travelled for days at a time whereas nowadays they usually make day trips during which they do not go very far out onto the drift ice. The same route was mentioned by another informant from Innaarsuit, who indicated that during hunting trips from boats the island can be used as lookout posts from which hunters can scout for bears. A couple of informants from Naajaat and Tussaaq said that their routes take a westerly direction, adding that they are of course adjusted according to the ice conditions. In this area, a hunter (aged 53) also mentioned that prior to the early 1980s it was unusual to bring home a bear after a day's hunting trip where as now it had become more common.

The same hunting pattern was found in Aappilattoq and the town of Upernavik, where a number of hunters mentioned they travel west to the areas of shallow water in the spring. Six hunters from these two towns said that they did not have fixed hunting routes and that ice conditions determine the route taken. Two of these informants, who came from Upernavik town, said that hunters travel by boat to the west towards the "open sea" specifically to hunt bears. A third informant described the situation in the following statement:

> When we are hunting, we find bear tracks no matter where we go. We sail to where we think the bears are, you know? We sail slightly to the west and then even further to the west. Then we find tracks. And we also overlook bears that are lying down ("*nullangasut*" = bears that are lying down to digest their food; author's note), for example, on the pack ice or behind the ice hummocks … we don't see the ones that are hiding well inside the ice. Sometimes we hunt very superficially ... that is … huh..

6. Do a greater number of polar bears come to visit/come of their own accord?

Among the 29 respondents, there were about the same numbers who answered "yes" (n = 13; 45%) and "no" (12 = 41%), and 4 (14%) did not express any opinion on the subject (Table 76).

A 54-year-old informant from Nutaarmiut said that there had not been much talk about polar bears when he was growing up in the 1950s, but that since the early 1980s they have been observed more frequently. In Innaarsuit, a 68-year-old man said that the number of bears has increased and that they come in the autumn after the polar night has set in where it can be dangerous to move around in the neighbourhood even just a few kilometres from the settlement. Another informant, from Nutaarmiut, said that these bears are hungry. A 53-year-old informant from Tussaaq was of the opinion that the increase in the number of bear visits has taken place during the last 20 years and that the visiting bears are primarily old animals with broken teeth or newly weaned young animals that find it difficult to fend for themselves. A 46-year-old hunter from Aappilattoq suggested the explanation that the change in occurrence is perhaps a result of an increase in temperature.

Six of the seven respondents from Upernavik answered this question in the affirmative. One of them, a 42-year-old hunter said that in the late 1990s and the early 2000s many more bears have come into the area, and another said that the bears can be seen from the town. A third hunter from the same town said that more bears are now observed in the summer.

However, the data on the individual bear catches in the area indicate that the number of polar bears that come into inhabited areas of their own accord has remained consistently small (see "Hunting methods" in "The catch in figures", p. 196).

	Total	Nutaar-miut	Innaar-suit	Naajaat, Tussaaq Aappilattoq	Upernavik town
Yes	13	1	2	4	6
No/Same	12	2	5	4	1
No opinion	4	-	2	2	-
N_{total} (hunters asked)	29	3	9	10	7

Table 76. Distribution of responses to question 6, from the area to the south of Nuussuaq (Area 4) in the Upernavik municipality: "Do a greater number of polar bears come to visit/come of their own accord?"

Driving a dog sled in the municipality of Upernavik. Photo: A. Heilmann

7. Have your routes changed in recent years?

A total of 23 hunters were asked this question, of which 8 (ca. 35%) answered "yes", 13 (ca. 56%) answered "no" and 2 (ca. 9%) had no specific opinion on this matter (Table 77). There was however no statistically significant difference in the numbers of affirmative and negative responses ($\chi^2 = 0.589$, $P = 0.443$, $df = 1$) or in the age of the respondents in the two groups ($t = 1.239$, $P = 0.231$, $df=19$).

Four of the eight respondents who said "yes" indicated that the routes had changed in recent years because the ice has "gotten worse". These four informants lived in Nutaarmiut, Innaarsuit, Aappilattoq and Upernavik town, respectively. A hunter from Nutaarmiut described how in 2005, it had been difficult to pass over the leads which run parallel to the coast in order to drive out to the west on to the drift ice because the leads did not freeze over (see Fig. 25). According to three other

	Total	Nutaar-miut	Innaar-suit	Naajaat, Tussaaq Aappilattoq	Upernavik town
Yes	8	1	3	1	3
No/Same	13	1	4	6	2
No opinion	2	-	-	2	-
N_{total} (hunters asked)	23	2	7	9	5

Table 77. Distribution of responses to question 7, from the area to the south of Nuussuaq (Area 4) in the Upernavik municipality: "Have your routes changed in recent years?"

informants (2 in Innaarsuit and 1 in Upernavik town) the changes to routes of travel are especially due to the fact that these cracks in the ice to which the polar bears are attracted are now situated closer to the coast. One other informant did not give any explanation for his response.

8. *Are a greater number of polar bears caught from boats than previously?*

A total of 28 polar bear hunters were asked this question, of which 25 (ca. 89%) answered "yes" and 3 (11%) had no particular opinion on the subject (Table 78).

	Total	Nutaar-miut	Innaar-suit	Naajaat, Tussaaq Aappilattoq	Upernavik town
Yes	25	2	7	9	7
No opinion	3	1	2	-	-
N_{total} (hunters asked)	28	3	9	9	7

Table 78. Distribution of responses to question 8, from the area to the south of Nuussuaq (Area 4) in the Upernavik municipality: "Are a greater number of polar bears caught from boats than previously?"

In Nutaarmiut a 45-year-old man, who is a particularly active polar bear hunter and who only uses a dogsled when hunting polar bears, said:

> From a boat...? (sighs) That is correct ... it's true. There are more of them. Perhaps because they have now learned how. They are learning. Especially in the spring.

He implied in the above statement that a number of hunters in the town have started to learn how to catch polar bears, particularly because it can take place from a boat. Another hunter from the same place said that in 2002 and 2003, many bears were caught from cutters "out west" (*i.e.* among the drift ice in Baffin Bay).

A 73-year-old hunter from Innaarsuit remarked that the number of hunters has increased and that they are more effective because of the use of motorised boats:

> These days there are many more hunters here. There are lots of them now. When we were kids (*i.e.* in the 1940s; author's note), when we started to hunt, there were just 4 hunters. Now there are lots of hunters in Innaarsuit. And they don't even need to look for bears properly. They just wait until they run into the bears. In the year 2005 I don't even know how many people caught bears. There

> were so many [........]. It was not like that in the past. [........] we are about to have open sea (ice breakup in spring; author's note), so the boats with motors go westwards. These boats come back with a lot of bears. Ila ila... They can catch a lot of bears very quickly.

Another hunter, who was almost 70 years old, explained that especially in the spring, the bears are caught from boats. A 53-year-old informant from Tussaaq voiced his opinion as follows:

> Yes ... that is, bears caught from skiffs. The skiffs have become much too fast, nothing is too far away for them anymore. That is my assessment, and also that of those who I have spoke to, they say the same thing.

The same man also stated that the polar bears gather around seal carcasses that are left on the drift ice by certain hunters and that in this way the bears have become an easy prey that has been caught in great numbers in recent years. On the subject of the use of boats as opposed to dogsleds, he said:

> They set sail in the daytime, before it gets light – the next day they arrive with *pingajoqqat* (female with two cubs; author's note). With the mother. Two-year-olds ... close to here. They are called *pingajoqqat*. Some people hunt them one at a time. Some of them ... in the course of one month, maybe some have talked about this ... sometimes during one month they catch more than five, they bring about 6 home. I would evaluate this sort of thing as an attempt at extermination. As I said. But hunting from dogsleds does not threaten the population.

In Aappilattoq, all of the seven informants who were asked this question answered it in the affirmative. A 50-year-old hunter said:

> The fact that the weather has got warmer has had the effect that the skiffs ... they come out to the leads out there and sail on the open water. There are also narwhals in the leads. There are lots of game animals there. Loads and loads of seals, loads of birds out there and lots of bears. It seems as though there are lots of bears.

Another hunter expressed the same idea in the following statement:

> When they set sail, one might say, that when they sail a long way out, it is maybe a sure thing that they catch bears.

All of the seven informants from Upernavik town expressed the opinion that the number of bears caught from boats has increased in recent years. One informant specified that this increase has taken place since 1996–1997. In this area, too, some informants said that the increase is due to the light ice conditions especially during the spring and that the occurrence of bears has increased.

The data about the individual bear catches clearly indicates that there has been an increase in the proportion of bears that are shot from boats in this part of the Upernavik municipality (see the answers to question 25, and the section "Hunting methods" in "The catch in figures", p. 196).

9. Are a greater number of polar bears caught on land than previously?

A total of 23 informants were asked this question, 5 of which (22%) answered "yes", 11 (48%) answered "no" and 7 (30%) had no opinion on the subject (Table 79).

Among the respondents who answered in the affirmative, there were two who elaborated on their responses. A 45-year-old hunter from Tussaaq was of the opinion that the polar bears now occur more frequently on land, partly because they seek refuge in the case of poor ice conditions and stormy weather. However, he considered this to be normal and added that the bears' condition is good. A 27-year-old from Innaarsuit suggested that more bears are caught on land because they now move around more and because they have increased in number and therefore occur more frequently along the shore.

	Total	Nutaarmiut	Innaarsuit	Naajaat, Tussaaq Aappilattoq	Upernavik town
Yes	5	1	2	1	1
No/Same	11	1	5	4	1
No opinion	7	-	1	3	3
N_{total} (hunters asked)	23	2	8	8	5

Table 79. Distribution of responses to question 9, from the area to the south of Nuussuaq (Area 4) in the Upernavik municipality: "Are a greater number of polar bears caught on land than previously?"

Occurrence of polar bears and dens

10. Do the bears occur in particular areas?

Nineteen (68%) of the 28 respondents to this question answered "yes", 6 (21%) answered "no" and 3 (11%) had no opinion on the subject (Table 80).

	Total	Nutaar-miut	Innaar-suit	Naajaat, Tussaaq Aappilattoq	Upernavik town
Yes	19	1	6	6	6
No	6	1	2	2	1
No opinion	3	-	1	2	-
N_{total} (hunters asked)	28	2	9	10	7

Table 80. Distribution of responses to question 10, from the area to the south of Nuussuaq (Area 4) in the Upernavik municipality: "Do the bears occur in particular areas?"

The general statements that were made indicated that the bears occur (1) on the offshore drift ice and (2) at the leads which run parallel to the shore (*i.e.* in the shear-zone between the land fast ice and the drift ice (Table 81).

	Number of responses	Nutaar-miut	Innaar-suit	Naajaat, Tussaaq Aappilattoq	Upernavik town
The drift ice "in the West" (*i.e.* Baffin Bay)	8	1	1	3	3
Channels and cracks in the ice	7	1	4	1	1
Other places	9	-	1	4	4
N_{total}(responses)	24	2	6	8	8

Table 81. Types of responses to question 10, from the area to the south of Nuussuaq (Area 4) in the Upernavik municipality: "Do the bears occur in particular areas?"

Among the informants who were of the opinion that the bears prefer the drift ice, there was one hunter from Nutaarmiut who said that between February and April, the bears occur at the eastern edge of the Baffin Bay pack ice where there are three parallel areas of shallow water off the coast (probably the areas where the leads which other informants have described are located; author's note), characterised by stranded icebergs (Fig. 25). Several informants mentioned these leads (*aalaneq/aalanerit*) that run along the coastline. A hunter from Aappilattoq spoke of three leads which attract bears because they seek food there. The same phenomenon was described by a hunter from Upernavik town, who referred to the area outside of the town of Upernavik as *aalaneqarfiit* (a place where there are openings in the ice because of, for example, sea currents).

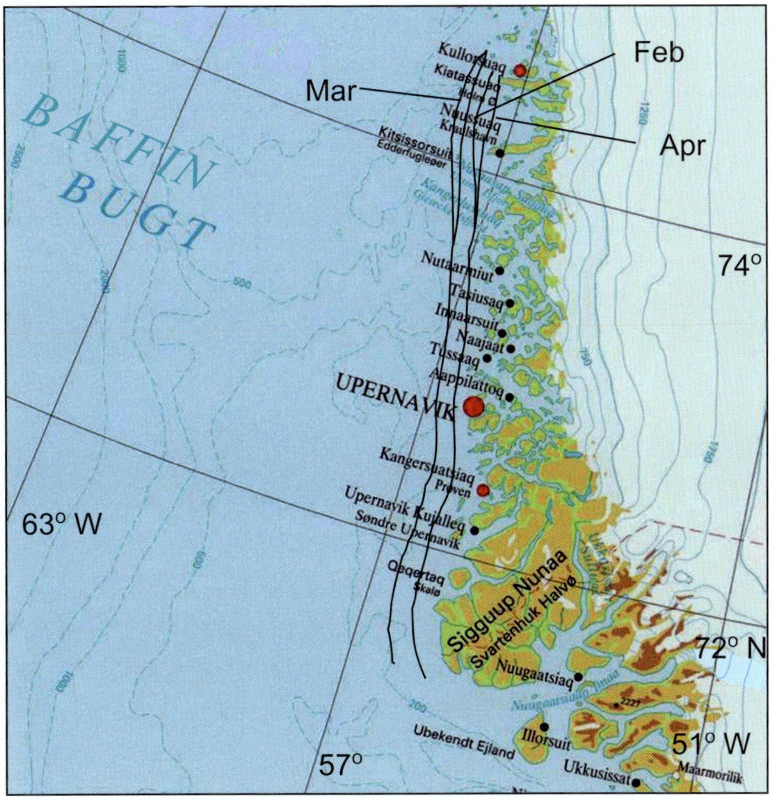

FIG. 25. Lead systems in the sea ice along the land, as drawn by a hunter from Nutaarmiut, during the interviews in the Upernavik municipality (Area 4) in February 2006.

The responses from different informants to this question included:

> The bears are maybe in the area to the south of Upernavik town, and around the route to Naajaat. In the spring, they may be encountered just as they leave their dens. They are also caught, for example, around Qaarsorsuaq...

> There are a large number of seals to the west of Upernavik town. Out there, the bears reign ... one might say, out on the ice. That is their hunting ground.

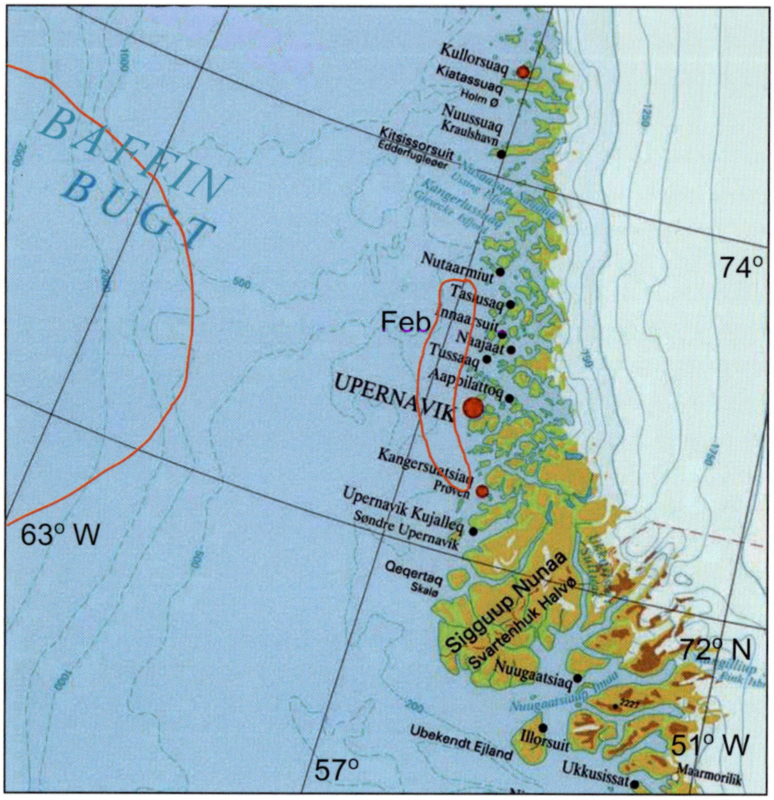

FIG. 26.
The general winter and spring distribution of polar bears as drawn on a map by two hunters from Aappilattoq and Upernavik town during the interviews in the southern part of the Upernavik municipality (Area 4) in February 2006.

> It is on the sea ice and also on land that they build dens in the three months, during which I have heard that the bears do not eat anything.
>
> From the area here, from north to south. They go to this islands (*i.e.* the islands to the west of Innaarsuit; author's note). To the south. And when it starts to get light, then they go west.

A couple of hunters identified some areas where bears are frequently encountered on maps (Fig. 26).

11. Where have you seen bear tracks?

All of the 28 informants talked about areas where they had seen tracks (1 hunter from Nutaarmiut was not asked). Unsurprisingly, tracks had been observed in a number of areas and several locations were mentioned: In between the islands Kingittuarsuk and Kingittortallit, in the vicinity of Tussaaq, Kuutseqarfeertorraanni by Innaarsuit

FIG. 27.
Areas where bear tracks (red lines) are often observed, as drawn by eight hunters during the interviews in the southern part of the Upernavik municipality (Area 4) in February 2006. Green arrow: Tracks made by females with cubs, in the spring.

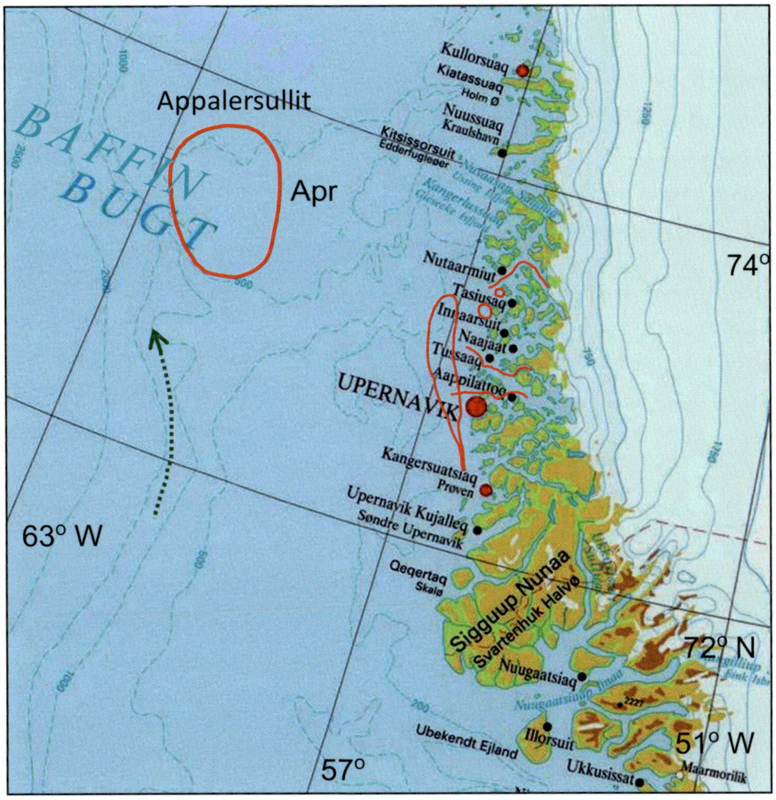

where Greenland halibut are fished, Eqqersoq and to the west of Upernaviarsuk close to Nutaarmiut. The majority of the responses were about observations of bear tracks on the ice to the west of populated areas (Fig. 27). A 53-year-old hunter from Tussaaq was of the opinion that the tracks have "come closer" (to land and populated areas; author's note) since the early 1980s. The tendency of the tracks to now occur closer to the coast was also mentioned by other informants. A hunter from Upernavik said that previously (*i.e.* in the late 1980s) hunters would only reach areas where there were polar bear tracks after having driven a dogsled west over the drift ice for three days to an area of shallow water.

There were different opinions as to the general direction of the tracks, but the majority of the responses indicated that in the spring they tend to go from north to south and from south to north. This may reflect the fact that the bears mainly follow the aforementioned shore leads. One informant stated that the tracks particularly seem to indicate a northerly direction of travel when the ice forms in the autumn. An 18-year-old hunter from Upernavik town had noticed that the tracks went towards the shore at night and to the west again at dawn, but a 43-year-old hunter from the same town said that the direction of the tracks is determined by the wind direction.

A 40-year-old informant from Upernavik said that the polar bears (in March?; author's note) move towards the leads when the currents get stronger according to the lunar phases. According to him, there is a tendency towards a difference in the distribution pattern at the three leads, so that females with cubs move along the easternmost lead whereas the tracks of larger bears (males?; author's note) occur along the more offshore lead.

A 49-year-old hunter from Upernavik town reported having observed a large number of tracks on the drift ice (Fig. 27):

> ... I go to, from Upernavik and to the west, the shallow water that is here, *Appalersullit*, that is out to the west, a large bear area [... Fig. 27] There were loads of bears here in April. I came across masses of tracks (in April 1988), after having travelled over some ice [...] Then there was a broad *illineq* – a path over the snow (*illineq* – trails on the snow; this term is mostly used when talking about dogsled trails; author's note). A very big path in the snow – the bear's path. It was maybe during the mating season, when they had followed each other's tracks.

A 50-year-old hunter from Aappilattoq spoke of some enormous tracks on the drift ice:

> When we were out there, a long, log way out there (after having driven) for three days. There where we could no longer see land. The mountains around Upernavik town are quite high. Some of them are 1 km high, others are 700 m high. If you want to see bear tracks, you have to be a long way out. Perhaps 300 km to the west of the shore [........] One spring, in one single day, we came across tracks made by bear number 17 (*i.e.* the 17th bear observed during that particular trip; author's note). I measured them. It must be a giant bear, that one. I had measured the handle of my whip ... how long is my arm ... 20 ... and halfway here. From my hand to here. Just the sole. The bear's sole. [........] Maybe about 35–40 cm wide. You know? The sole was about like that. And the foot, you know ... It must be an enormous bear. Some say that some bears are so big that there are chunks of ice in their fur. When they have been in their house (den), or when they resurface after swimming, then the ice forms in the back part of their fur. These are called *sermertalissuit* – the ones with chunks of ice in their fur.

12. Do the bears have regular migration routes?

There were divided opinions on this matter. A total of 27 hunters were asked this question: 13 (48%) answered "yes", 10 (37%) answered "no" and 4 (15%) had no opinion on the matter (Table 82).

	Total	Nutaar-miut	Innaar-suit	Naajaat, Tussaaq Aappilattoq	Upernavik town
Yes	13	2	4	6	1
No	10	-	5	1	4
No opinion	4	-	-	3	1
N_{total} (hunters asked)	27	2	9	10	6

Table 82. Distribution of responses to question 12, from the area to the south of Nuussuaq (Area 4) in the Upernavik municipality: "Do the bears have regular migration routes?"

The affirmative responses indicated that the bears migrate (1) along the leads in the ice along the shore (n = 8 answers), (2) closer to the land when the ice is forming because they seek areas where there is thin ice (n = 2), (3) in the fjords (n = 1), and (4) in a northerly direction in the winter (n = 1). One respondent did not give any explanation for his answer. These responses thus indicated that there was not any general opinion that the bears have specific migration patterns. This was also implied by the remaining responses, which suggested that the bears' migration is determined by the ice conditions and by their sense of smell (*i.e.* the direction of the wind; author's note). One informant was of the opinion that the bears mainly move to the east when the ice is stable and to the west when it is unstable. This perception was shared by another informant, who thought that the polar bears are primarily to be found on the drift ice in Baffin Bay where they follow the retreat of the (melting) ice towards Canada in the spring, and once again move eastwards with the ice towards Greenland in the autumn, when the ice forms.

13. Have you seen very small bear tracks (i.e. from cubs of the year)?

Among the 27 respondents who were asked this question, there were 8 (31%) hunters who had seen this kind of track, 18 (65%) had not, and 1 (4%) did not express any opinion on this matter (Table 83).

Five informants stated that they had seen very small tracks (*i.e.* made by cubs that had just left the den, or by cubs that are less than 1 year of age) out on the drift ice to the west, and two reported having observed such tracks in Qimusseriarsuaq/

	Total	Nutaar-miut	Innaar-suit	Naajaat, Tussaaq Aappilattoq	Upernavik town
Yes	8	1	4	2	1
No	18	1	5	8	4
No opinion	1	-	-	-	1
N_{total} (hunters asked)	27	2	9	10	6

Table 83. Distribution of responses to question 13, from the area to the south of Nuussuaq (Area 4) in the Upernavik municipality: "Have you seen very small bear tracks?"

Melville Bay (Figs. 21, 27). These observations were made in April–June and, according to the hunters, the cubs come out of the den from April to early May. The general absence of observations of tracks made by females with newborn cubs indicates that not many bears go into dens to give birth to cubs in this part of the municipality of Upernavik.

14. Where do the bears spend the summer?

This question was put to 28 hunters, 13 (46%) of whom expressed an opinion on the subject and 15 (54%) of whom had no particular opinion on the subject (Table 84).

	Total	Nutaar-miut	Innaar-suit	Naajaat, Tussaaq Aappilattoq	Upernavik town
Opinion on this	13	1	4	5	3
No opinion	15	1	5	5	4
N_{total} (hunters asked)	28	2	9	10	7

Table 84. Distribution of responses to question 14, from the area to the south of Nuussuaq (Area 4) in the Upernavik municipality: "Where do the bears spend the summer?"

	Number of responses	Nutaar-miut	Innaar-suit	Naajaat, Tussaaq Aappilattoq	Upernavik town
On the drift ice	6	-	2	3	1
On land	2	1	-	1	-
Drift ice and on land	3	-	2	1	-
On the glaciers	3	-	1	1	1
Canada	1	-	-	-	1
N_{total} (responses)	15	1	5	6	3

Table 85. Types of responses to question 14, from the area to the south of Nuussuaq (Area 4) in the Upernavik municipality: "Where do the bears spend the summer?"

The respondents who had an opinion on the subject made various suggestions as to where the bears are to be found in the summer (Table 85). In the Qaanaaq municipality and in the northern part of the Upernavik municipality, several of the informants said that they did not really know where the bears go in the summer and the responses indicated on the whole that it is unusual to see bears at that time of year. The majority of the informants who had an opinion regarding the whereabouts of the bears in the summer thought that they are to be found on the drift ice in Baffin Bay or on land. A couple of informants specified that these animals stay on the drift ice and as it melts they move gradually towards Canada. One of the three informants who were of the opinion that the bears are to be found on or around the glaciers in the summer said that this was true of the glaciers in Qimusseriarsuaq / Melville Bay.

A 49-year-old hunter from Upernavik town, who was one of the respondents who did not have any opinion regarding the bears' whereabouts in the summer, said that he had never shot a bear in the summer. A 43-year-old hunter from Upernavik and a 50-year-old from Aappilattoq simply said that the bears are not seen during the summer.

FIG. 28. Observations of maternity and temporary dens, as indicated on a map by 18 hunters during the interviews in the southern part of the Upernavik municipality (Area 4) in February 2006. Orange triangle = temporary dens; green dots = maternity dens; ? = year of observation was not specified.

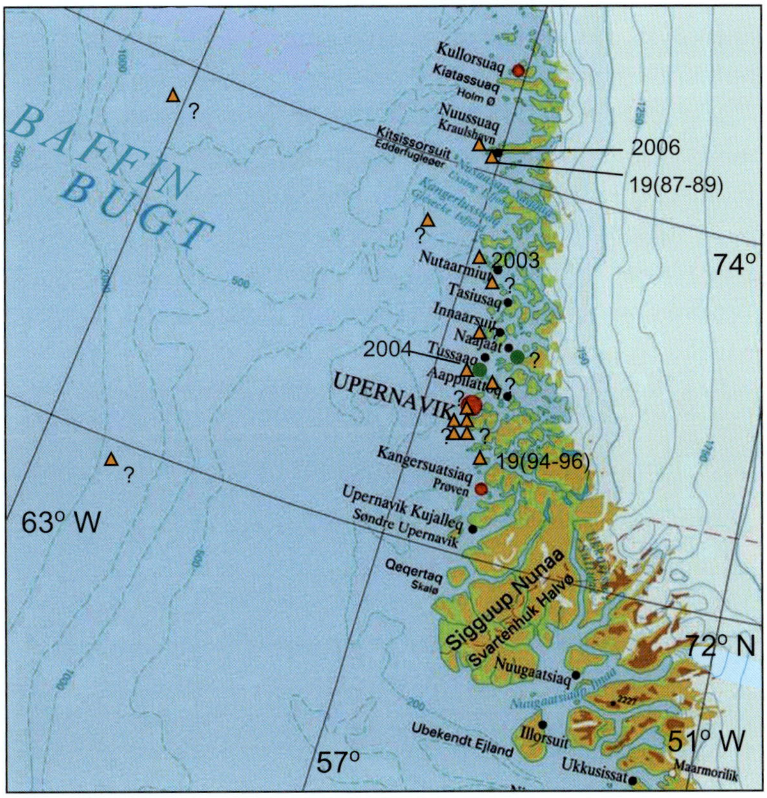

15. Have you seen dens with females with small cubs (maternity dens)?

Only one of the 28 respondents to this question had seen a maternity den, but one other informant mentioned a den that had been observed by a third party. An 18-year-old hunter from Upernavik said that he had seen an abandoned den from a skiff in February 2004. The den was positioned "not that high up" in the mountain on the island of Kingiitortataalliit, to the west of the mouth of the Upernavik Isfjord. Another hunter said that he had heard about some hunters from other settlements who had shot a family of bears in a den high up in the mountains by Sisorarsuit, east of Tussaaq in 1993 (Fig. 28). The remarkably small number of observations of maternity dens, along with the small number of observations of tracks made by young cubs, indicates that few females choose to give birth to their cubs in this part of the municipality.

16. Have you seen temporary dens?

Sixteen (59%) out of 27 respondents had seen temporary dens, whereas 11 (41%) had not (Table 86). However, for five of the dens, it was the case that several different hunters talked about the same den. The occurrence of temporary dens in the southern part of the Upernavik municipality is shown in Fig. 28.

Some of the respondents described temporary dens in detail. According to one hunter from Nutaarmiut, the bears dig in the snow, for example next to an iceberg, to make a den while they are moving about on the drift ice. Another hunter from the same settlement mentioned a case where he had followed the fresh tracks of a young male, which had dug itself a den near to Toqqusaq in February (or March; year unknown). After having dug out an opening and shooting the bear, they noted that the entrance was 50 cm wide and the den itself was a couple of metres long and less than circa 1 m high and had no ventilation hole.

A report from Innaarsuit described a female with two relatively large cubs (possibly more than 1 year old; author's note) which were tracked to a den in the

	Total	Nutaar-miut	Innaar-suit	Naajaat, Tussaaq Aappilattoq	Upernavik town
Yes	16	2	6	4	4
No	11	1	3	4	3
N_{total} (hunters asked)	27	3	9	8	7

Table 86. Distribution of responses to question 16, from the area to the south of Nuussuaq (Area 4) in the Upernavik municipality: "Have you seen temporary dens?"

mountains in March 2001. The hunters had lost contact with the bears the previous day and had fled up a steep rock face, where they had later that night dug a den in the hard snow. The bears were shot in their den. A similar case, where a single bear was shot in its den at Nuna avalernani in February 2006, was recounted by a hunter from Innaarsuit. The bear had apparently been in the den since the autumn of 2005.

A hunter from Tussaaq reported that the hunters from Kullorsuaq and Nuussuaq had told him that the bears sometimes build dens on the glaciers on Kiatassuit during the summer. The same informant reported having observed a temporary den on the drift ice (*maniilarsuit*), where a bear had rested for a while (*nullangasoq*) during bad weather, and where it had also shed some fur.

An informant from Upernavik had examined a temporary den that was over 1 m high and perhaps 2 m deep. Another hunter from the same town described a den that was 1 m high and around 3–4 m deep, which had been used by a lone bear.

17. Have there been changes to the occurrence of dens over the years?

A total of 18 hunters were asked this question, 1 of whom (6%) answered "yes", 2 of whom (12%) said "no" and 15 (82%) of whom had no opinion on the subject (Table 87).

There is frequently open water just outside of the town of Upernavik (population: 1099 in 2005) in early spring. Photo: E.W. Born

	Total	Nutaar-miut	Innaar-suit	Naajaat, Tussaaq Aappilattoq	Upernavik town
Yes	1	-	-	-	1
No	2	1	1	-	-
No opinion	15	1	5	5	4
N_{total} (hunters asked)	18	2	6	5	5

Table 87. Distribution of responses to question 17, from the area to the south of Nuussuaq (Area 4) in the Upernavik municipality: "Have there been changes to the occurrence of dens over the years?"

Considering how few dens had been observed in this area, it was not surprising that the majority of the respondents did not express any opinion when asked this question. However, one response given by a 49-year-old hunter from Upernavik implied that the number of dens has decreased because of changes to snow and ice conditions:

> Nowadays, it does not snow as much anymore. Snow is now in short supply on land. The large masses of snow and the ice come so late. Back when the ice used to come early, in November–December, the bears used to come onto land while it was still dark. There were more of them then (dens; author's note). Nowadays the ice comes after the light returns, you know, in January–February. All of these things take place later now. Everything is changing.

A 40-year-old hunter from Innaarsuit shared the same opinion as the informant who made the statement above, although he generally tended towards the belief that the occurrence of dens was unchanged:

> Don't know. Maybe it is unchanged. It's also the snow, isn't it? In recent years the snowfall has been less frequent. In some places, for example last year, you know. It rained a lot then, and the weather was bad. This year the weather has been very stable. We have had very fine weather, and at times the snow has been very good. They (dens; author's note) are probably to be found in some places – this year too.

Climate change

18. Have you observed changes to ice conditions?

Of the 28 respondents to this question, 26 (93%) answered "yes", whereas 2 (7%) did not express any opinion on the matter (Table 88).

Twenty-two (85%) of the respondents stated that the ice forms later and breaks up earlier and has become thinner and less safe to travel on. It was also evident from the responses that the ice has become more unstable and that there are more and larger holes in it (Table 89).

A hunter from Nutaarmiut remarked that since 2003, there has only been a little ice and that the ice has not been forming until January–February. This was confirmed by a 54-year-old who said that the ice used to form in December when he was a child (*i.e.* in the 1960s; author's note). The same informant mentioned that the ice has also got thinner compared to the past.

Six of the responses given by informants from Innaarsuit also indicated the same pattern. In this settlement a 68-year-old hunter said that in 2000, 2002, and 2003 the ice had not formed properly until March, and that during his childhood

	Total	Nutaar-miut	Innaar-suit	Naajaat, Tussaaq Aappilattoq	Upernavik town
Yes	26	3	8	9	6
No	2	-	1	1	-
N_{total} (hunters asked)	28	3	9	10	6

Table 88. Distribution of responses to question 18, from the area to the south of Nuussuaq (Area 4) in the Upernavik municipality: "Have you observed changes to ice conditions?"

	Number of responses	Nutaar-miut	Innaar-suit	Naajaat, Tussaaq Aappilattoq	Upernavik town
Forms later and breaks up earlier	14	2	6	3	3
Thinner and more unsafe	8	-	2	4	2
Holes in the ice and more open water	1	-	-	1	-
Changes to the ice edges	1	1	-	-	-
Other	2	-	-	1	1
N_{total} (responses)	26	3	8	9	6

Table 89. Types of responses to question 18, from the area to the south of Nuussuaq (Area 4) in the Upernavik municipality: "Have you observed changes to ice conditions?"

it generally used to form in January. Two other informants were of the opinion that in recent years the ice had not formed until January, whereas this used to take place in November–December. One of these hunters specified that 2003 and 2004 had been years with very little ice and was of the opinion that this was due to an increase in temperature.

In Aappilattoq and the town of Upernavik too, informants spoke of how in recent years the ice has been forming later and breaking up earlier (Table 89). Another phenomenon which was mentioned by several informants is that the ice has become more unstable and can break up several times during the course of the winter. One 43-year-old hunter from Aappilattoq described it as follows:

> In recent years there have been big changes. Last year, the sea ice was also like it is now, but the ice broke up several times. We had open water maybe four times, and we didn't get ice for real until the beginning of February. This year (2006; author's note) we got ice before we got very far into January.

An informant from Nuussuaq spoke of how there had previously been unsafe ice and holes in the ice by Sugar Loaf Island, but that nowadays this kind of ice occurs slightly further to the east by an island called Timilerssua. A hunter from Innaarsuit reported that the same kind of ice occurs around the islands of Kitsissut and Kingittuarsuk, to the west and northwest of Tussaaq respectively.

However, several hunters described how the ice conditions in the winter of 2005/2006 had been safer, and according to one informant, the ice had already formed in December.

Other comments were made regarding changes to the ice conditions. A 75-year-old hunter from Nutaarmiut gave the following response:

> [........] The weather is getting milder all the time. And I have heard more than just one hunter saying: "My nets are on land!" There is no ice shelf (ice foot); the ice comes without forming *qaanngut* (*i.e.* a shelf of ice attached to the coast). When the ice comes and the hunters set their nets where they usually do ... then they are set on land. The nets are otherwise supposed to be attached to the ice foot. That's how much things have changed.

A 40-year-old hunter from Naajaat stated that it was obvious that the ice had changed a great deal in the areas close to the coast. He had also noticed that the heavy drift ice in Baffin Bay is now found further and further to the west, away from the coast.

A 42- and 43-year-old hunter from Upernavik town who were interviewed together both stated:

> From 2001 to 2005, it was very difficult for the sea to freeze over. The ice came and melted again. The weather got milder repeatedly. It was very mild. January, February, March. They were very mild months. It kept getting milder after the frost had set in ... again and again during that period. But this year – 2006 – the frost is like it used to be. But it is still not as cold as it was in the past.

In response to the question about the extent to which ice conditions have changed, a 49-year-old hunter from Upernavik town said:

> Have they ever! They have changed a lot. The year before last (2004?; author's note), when the ice came, we caught a load of bears in the Upernavik municipality. Perhaps more that 30 bears were caught, from skiffs alone. And when the helicopters fly to Kullorsuaq, they have seen more than 40 bears here. By the edge of the ice, you know. We just sail towards them in skiffs.

19. Have you seen changes to the icebergs?

Nine (ca. 32%) of the 28 hunters who were asked this question answered "yes", 16 (57%) answered "no", or that there had been no change, and 3 (11%) had no opinion on the subject (Table 90).

Among those informants who had observed changes, the responses were more or less equally divided between those who said that the number of icebergs had increased and those who said that the number had decreased (Table 91).

A hunter from Nutaarmiut had noted that there had not been any icebergs "this year" (the winter of 2005/06; author's note) around Tasiusaq and to the west of

	Total	Nutaar-miut	Innaar-suit	Naajaat, Tussaaq Aappilattoq	Upernavik town
Yes	9	1	2	5	1
No/Same	16	1	6	4	5
No opinion	3	1	1	1	-
N_{total} (hunters asked)	28	3	9	10	6

Table 90. Distribution of responses to question 19, from the area to the south of Nuussuaq (Area 4) in the Upernavik municipality: "Have you seen changes to the icebergs?"

	Number of responses	Nutaar-miut	Innaar-suit	Naajaat, Tussaaq Aappilattoq	Upernavik town
Decreased in number	5	1	2	2	-
Increased in number	4	-	-	3	1
N_{total} (responses)	9	1	2	5	1

Table 91. Types of responses to question 19, from the area to the south of Nuussuaq (Area 4) in the Upernavik municipality: "Have you seen changes to the icebergs?"

Nutaarmiut, and said that this was perhaps a result of the bad weather. He added that hunters usually set their seal nets at the icebergs.

A 42-year-old from Innaarsuit stated that icebergs had become scarce in the area, whereas in the 1960s and 1970s the icebergs had come close to the settlement. A 68-year-old informant from the same settlement had noticed that there are no longer any stranded icebergs in the area to the east of the settlement in the autumn, probably because of the increased currents. Such icebergs had previously provided shelter from the wind for hunters who were out hunting seals in the autumn.

In response to this question, a 50-year-old hunter from Aappilattoq said:

> When we were kids, we fetched ice from the beach; ice on land (*nilak* – freshwater ice) [........] There is still some *nilak*, but not as much as there used to be. In about the 1970s it was not possible to reach Aappilattoq for the whole month of July, I can tell you. When the ice comes (*sikkeraangat* in the Upernavik dialect; *sikugaangat* in West Greenlandic), the boats turn back when they reach the islands out there. They are not able to start coming here until the end of July or during August. KNI's (the trade company) cargo ships.

Four hunters were of the opinion that the number of icebergs had increased (Table 91). One of these informants, a 47-year-old man from Aappilattoq gave a response which indicated that the icebergs move around more, which may mean a greater concentration of icebergs in some areas:

> Square-shaped icebergs ... we call them *maniitsut* – the large, high icebergs with columns. I have noticed that the glacier has moved to the east. Many people can testify to this. That area, where we can't fish ... but this summer people have begun to fish in a very strange area. This is because the icebergs have sailed away from here. There are no longer any icebergs that stand still ... it is obvious.

20. Have you seen any changes to the glaciers?

Twenty-five (89%) of the 28 informants who were asked this question answered "yes", one (4%) answered "no" and two (7%) had no specific opinion on the matter (Table 92).

One response that was given by many of the informants (89%) was that the glaciers have clearly receded – *i.e.* in an easterly direction. Two respondents also informed the interviewer that the retreating glacial ice has exposed land and nunataks (*i.e.* areas of the mountains that are surrounded by glacier ice). Three respondents indicated that certain glaciers have increased in productivity in recent years (Table 93).

Two hunters from Innaarsuit specified that the glacier by Alianaatsorsuaq (to the east of Akkarnersuaq, the island to the east of Tasiussaq) had diminished, creating more straits which run a long way inland. In this area, the drifting icebergs have nearly all disappeared because of the current and the area is no longer unsafe for travel.

A 40-year-old hunter from Naajaat expressed the opinion that the glaciers in the area unmistakably calve earlier in the season because of the increase in the strength of the current. Another 40-year-old informant from Aappilattoq spoke of how the glaciers have started to move faster in the last 20 or so years. A 49-year-old man from Upernavik said that the glaciers calve more frequently nowadays.

	Total	Nutaar-miut	Innaar-suit	Naajaat, Tussaaq Aappilattoq	Upernavik town
Yes	25	2	8	10	5
No/Same	1	-	1	-	-
No opinion	2	1	-	-	1
N_{total} (hunters asked)	28	3	9	10	6

Table 92. Distribution of responses to question 20, from the area to the south of Nuussuaq (Area 4) in the Upernavik municipality: "Have you seen any changes to the glaciers?"

	Number of responses	Nutaar-miut	Innaar-suit	Naajaat, Tussaaq Aappilattoq	Upernavik town
Retreated	24	2	8	9	5
More productive	3	-	-	2	1
N_{total} (responses)	27	2	8	11	6

Table 93. Types of responses to question 20, from the area to the south of Nuussuaq (Area 4) in the Upernavik municipality: "Have you seen any changes to the glaciers?"

Two hunters, from Naajaat and Tussaaq respectively remarked that the receding glaciers have exposed land and that there are now new nunataks, examples of which have been observed between Qallersuaq and Tussaaq.

Several hunters talked about how the edge of the Upernavik Isstrøm ("Upernavik Ice Stream") has moved a long way to the east between Uilortussoq Island and the mountain area Akullikassak in the north.

21. Have you seen changes to the snow?

Eighteen (64%) of the 28 interviewees answered "yes" to this question, 8 (29%) answered "no", or that there had been no change and 2 (7%) had no opinion on the subject (Table 94).

The majority of the hunters who had noticed a change to the snowfall in recent years were of the opinion that there was less snow, and some of them had also noticed that it does not fall at the same times of year as it used to.

A few informants were however convinced that there has been a greater amount of snow in recent years (Table 95). One of these informants, a man from Innaarsuit, said that a lot of snow had fallen in 2004 and 2005. Another man from the same settlement said that there is a lot of snowfall in the autumn but that it disappears again when it rains. The idea that the heavy autumn snowfall disappears again was also expressed by a hunter from Upernavik town in the following statement:

	Total	Nutaar-miut	Innaar-suit	Naajaat, Tussaaq Aappilattoq	Upernavik town
Yes	18	2	6	6	4
No/Same	8	1	2	3	2
No opinion	2	-	1	1	-
N_{total} (hunters asked)	28	3	9	10	6

Table 94. Distribution of responses to question 21, from the area to the south of Nuussuaq (Area 4) in the Upernavik municipality: "Have you seen any changes to the snow?"

	Number of responses	Nutaar-miut	Innaar-suit	Naajaat, Tussaaq Aappilattoq	Upernavik town
Less falls	15	2	4	6	3
More falls	3	-	2	-	1
Change in the time of year	2	1	-	-	1
N_{total} (responses)	20	3	6	6	5

Table 95. Types of responses to question 21, from the area to the south of Nuussuaq (Area 4) in the Upernavik municipality: "Have you seen any changes to the snow?"

> For example, we get a lot of snowfall in the autumn now. Like in the autumn before last – in September – the snow came. That which otherwise would have been the winter snowfall, fell at the end of November and it was cold, but due to the very warm weather, all the snow melted. And later, after that … we get a lot of snow now too. Now, also in the spring, we get loads of snow.

A hunter from Nutaarmiut also expressed the opinion that there has recently, in contrast to in earlier periods, been a lot of snowfall in the spring:

> This year there is no snow (winter of 2005/06; author's note). We say that it is not until the spring comes [........] I told the others that it is because we only get snow when the spring comes. It has started to be like that in recent years.

However, on the whole, the responses gave the impression that the hunters have noted a smaller amount of snow in recent years and the snow that falls in the autumn tends to disappear because a thaw sets in accompanied, on occasion, by rainfall.

22. Have you seen any changes to current patterns?

A total of 28 hunters were asked this question, 21 (75%) answered "yes", four (14%) said that conditions were unchanged and three (11%) had no opinion on the matter (Table 96).

The majority of the respondents who elaborated on their response were of the opinion that the currents have increased in strength while in several cases it was also said that the currents have changed direction (Table 97).

Four informants (Innaarsuit: 2, Aappilattoq and Upernavik town) thought

	Total	Nutaar-miut	Innaar-suit	Naajaat, Tussaaq Aappilattoq	Upernavik town
Yes	21	3	6	8	4
No/Same	4	-	2	-	2
No opinion	3	-	1	2	-
N_{total} (hunters asked)	28	3	9	10	6

Table 96. Distribution of responses to question 22, from the area to the south of Nuussuaq (Area 4) in the Upernavik municipality: "Have you seen any changes to current patterns?"

	Number of responses	Nutaar-miut	Innaar-suit	Naajaat, Tussaaq Aappilattoq	Upernavik town
Have become stronger	13	3	3	4	3
Have changed direction	6	-	2	3	1
N_{total} (responses)	19	3	5	7	4

Table 97. Types of responses to question 22, from the area to the south of Nuussuaq (Area 4) in the Upernavik municipality: "Have you seen any changes to current patterns?"

that the currents have increased in strength because the glaciers have retreated. Two informants from Nutaarmiut and Upernavik town, respectively, said that the inshore current in particular has increased in strength whereas this does not seem to be the case for the offshore currents ("in the west"). Six interviewees (1 from Innaarsuit, 1 from Nutaarmiut, 1 from Tussaaq, 1 from Aappilattoq and 2 from Upernavik) emphasised that the stronger currents mean that the ice is thinner and cause more and larger holes in the ice. Six hunters (2 from Innaarsuit, 3 from Aappilattoq and 1 from Upernavik town) stated that the currents also seem to have changed direction in the fjords. A 48-year-old hunter from Upernavik commented on the sea temperature:

> The warm current lies like a duvet over our sea. That was the first thing I noticed, because I have been travelling since I was a child ... the way that the ice freezes over on the side of our boat. Since October 1999 ... 1998–1999, I have noticed that we get much less ice on the side of the boats. I have also noticed that the sea seems warmer. That is how it has become now. And it is still getting warmer.

One hunter who otherwise had no specific opinion on changes to the current patterns emphasised that the current is normally stronger when there is a full moon.

23. Have you seen any changes to the weather?

Of the 27 informants who were asked this question, 22 (ca. 81%) answered "yes", 4 (ca.15%) answered "no" and 1 (4%) had no particular opinion on the matter (Table 98).

As was the case in the Qaanaaq municipality and in the northern part of Upernavik municipality, hunters in the areas to the south of Nuussuaq had noticed

	Total	Nutaar-miut	Innaar-suit	Naajaat, Tussaaq Aappilattoq	Upernavik town
Yes	22	3	6	10	3
No / Same	4	-	2	-	2
No opinion	1	-	-	-	1
N_{total} (hunters asked)	27	3	8	10	6

Table 98. Distribution of responses to question 23, from the area to the south of Nuussuaq (Area 4) in the Upernavik municipality: "Have you seen any changes to the weather?"

	Number of responses	Nutaar-miut	Innaar-suit	Naajaat, Tussaaq Aappilattoq	Upernavik town
Warmer	7	1	2	4	-
More wind	7	-	1	3	3
Unstable / rainy	6	1	2	3	-
Other	3	1	1	1	-
N_{total} (responses)	23	3	6	11	3

Table 99. Types of responses to question 23, from the area to the south of Nuussuaq (Area 4) in the Upernavik municipality: "Have you seen any changes to the weather?"

that the temperature had increased and that the weather had become more stormy and unpredictable in recent years (Table 99).

Among those informants who said that the temperature increased, there was one hunter who said that the weather had got warmer compared to the "1900s". Another hunter said that the increase had occurred within the preceding five years. A 68-year-old hunter from Innaarsuit was however of the opinion that the conditions change cyclically:

> The change that I have noticed is that the frost is not able to form properly in the autumn. In the past, everything could freeze over as early as the middle of October, you know ... in the late 1980s. But because the ice conditions in the sea have been a bit changeable during the 10 year period up to 2000, with the ice sometimes forming very early ... then since the beginning of 2000, people tend to say that it is a result of climate change. But in my opinion, the weather changes over many years ... every tenth year the weather changed, so I don't think anything strange is going on. I also foresaw that we would have three years in 2002/03/04, when there would be almost no frost. And I imagine that in the next ten years, the sea ice will once again come at specific times ... that the sea ice will be there again.

Several of the responses to this question indicated that the weather has become stormier and that this has resulted in higher waves and generally more unsettled weather (Table 99). A number of respondents also said that the weather has on the whole become more humid with for example snowfall followed by rain in the autumn.

Other observations regarding the weather included (1) that the changes were observed in the last 2–3 years (2003–2005) but that the winter of 2005/06 was more "normal" and (2) that there has been snowfall in the spring when there is usually no snow.

24. Have you observed any other changes?

Seventeen (63%) of the 27 respondents to this question answered "yes", 8 (30%) answered "no" and 2 (7%) had no particular opinion on the subject (Table 100).

The comments regarding "other changes" involved polar bears, seals, walruses, narwhals, birds and other phenomena (Table 101). Although these responses are not directly related to the theme of "climate change", they will still be outlined in this section.

	Total	Nutaarmiut	Innaarsuit	Naajaat, Tussaaq Aappilattoq	Upernavik town
Yes	17	1	5	7	4
No	8	-	3	3	2
No opinion	2	2	-	-	-
N_{total} (hunters asked)	27	3	8	10	6

Table 100. Distribution of responses to question 24, from the area to the south of Nuussuaq (Area 4) in the Upernavik municipality: "Have you observed any other changes?"

Other Changes	Number of responses	Nutaarmiut	Innaarsuit	Naajaat, Tussaaq Aappilattoq	Upernavik town
Bears	3	-	2	-	1
Seals and walruses	8	1	3	3	1
Whales	5	-	1	1	3
Birds	4	-	-	4	-
Other	1	-	-	1	-
N_{total} (responses)	21	1	6	9	5

Table 101. Types of responses to question 24, from the area to the south of Nuussuaq (Area 4) in the Upernavik municipality: "Have you observed any other changes?"

Three hunters from Innaarsuit talked about polar bears. One of these hunters thought that bear tracks (*qallinikuusut*) had come closer, another was of the opinion that bears are sighted more frequently than they were previously, and the third man said that there were more bears (see also the responses to question 27).

Some of the responses also contained comments about seals. According to a hunter from Nutaarmiut and one from Innaarsuit the number of ringed seals has increased. An informant from Innaarsuit had noticed how the ringed seals that are caught in nets "are shedding their fur" (*mamaartut* – moulting), something which is usually only observed in the spring when the seals bask on the ice (*i.e.* as ringed seals are usually netted during winter darkness the statement indicates that in later years some seals seem to be moulting earlier in the season?; author's note). Another informant from the same settlement said that the number of seal pups (*torlusit*) has noticeably decreased – for information regarding the increased occurrence of large ringed seals see p. 120.

A hunter from Naajaat thought, in contrast to his colleagues from Nutaarmiut and Innaarsuit, that the ringed seals have become rarer. The ringed seals that, according to him, usually arrive in the coastal areas from the west and have not been seen in recent years. Nowadays, it is mainly harp seals that come in the autumn, as well as the "eastern seals" (*kangiata puisai* – which probably refers to the more "local" ringed seals found in the fjords to the east of Naajaat?; author's note).

Hunters mentioned changes to the occurrence of ringed seals in the Upernavik area. Photo: Ø. Wiig

Two hunters from Aappilattoq and one from Upernavik town conveyed that the ringed seals have occurred in greater numbers in recent years. A hunter from Upernavik suggested an explanation for the increased occurrence of bears in coastal areas is that the ringed seals now travel further east, perhaps in search of food or maybe because of changes to weather and ice conditions, and that they are therefore closer to the shore. He did however add that it may also be the result of an increase in the polar bear population.

An informant from Innaarsuit mentioned that walruses are no longer hunted, or that walrus hunting seldom takes place apparently because there is now less drift ice ("no longer any ice to the west of us").

Respondents also made comments about narwhals. A 66-year-old hunter from Innaarsuit thought that it was possible that the narwhals are fleeing from disturbance so that it seems as though they are "disappearing". He added that:

> For example narwhals. It is said that in southern Greenland there used to be many narwhals and that they are gone now. They are not gone. It is the traffic that runs day and night now … they flee from all that noise.

A 46-year-old informant from Aappilattoq said:

> May I mention these narwhals? It has been said that narwhals are decreasing in number, but there are loads of them. There are masses of them. I catch some every year.

An 18-year-old hunter from Upernavik said that he had noticed that the narwhals have chosen migration routes (this probably refers to the southerly migration in the autumn; author's note) that are noticeably further "offshore" – around the Upernavik municipality. This opinion was supported by a 49-year-old hunter from the same town, who thought that this was due to the disturbance caused by motorised skiffs in the spring and autumn. He explained that it is no longer possible to "stand and wait" (be on the lookout) at Kingittortalik, where he had otherwise caught many narwhals every spring. According to him, the narwhals flee under the ice as soon as they hear the sound of a motor, and in addition to this "all the paths have changed" (migration routes) because the weather and ice coverage have changed. A 48-year-old hunter from Upernavik mentioned that the narwhals no longer go into the fjords, staying instead in the offshore regions, and that they therefore have problems with getting to the places where they usually feed at that time of year (*i.e.* in the autumn).

Four hunters from Aappilattoq talked about birds. One hunter mentioned that the common eider is found in large numbers in the area, whereas there are fewer in other areas. While there had been fewer Brünnich's guillemots in recent years, they are starting to reappear. Another hunter was of the opinion that the guillemots had moved from south of the town of Upernavik and now occurred more frequently "to the north". On the other hand, the common eiders which used to occur in the areas to the east have moved to the islands in the west, since the eastern region has become a fishing area. Glaucous gulls (*Larus hyperboreus*) and cormorants (*Phalacrocorax carbo*) have begun to spend the winter "en masse" in the Aappilattoq area and the common eiders also overwinter there. A 50-year-old hunter said that the distribution of common eiders and Brünnich's guillemots had changed – "they have moved to other places" – from the Upernavik area to the bird cliff Apparsuit in the north. He also explained how during his childhood in spring, the men in kayaks laid in wait for passing common eiders and guillemots in the Nuugaarsuk cove slightly to the west of the oil tanks in Aappilattoq. When the kayaks were full, they rowed home. According to him, only a few eiders fly past that area nowadays. A 47-year-old hunter said that last year and the year before that (2004 and 2005; author's note) the common eiders had turned up in large flocks after the ice had formed. He had never seen this before and it was an unusual phenomenon during the period with sea ice. In January, when the holes in the ice were freezing over, they apparently headed towards areas of open water.

A hunter from Tussaaq remarked that the rivers have started to be extremely fast-flowing (probably because of the increased amount of rain in the summer; author's note).

Narwhals are found in leads outside of Aappilattoq. Photo: K. L. Laidre

Changes to the catch and occurrence of polar bears

25. Have the changes affected the polar bear hunt?

This question referred specifically to the extent to which climate-related changes to the physical environment (ice, wind, glaciers, weather etc.) have affected the bear hunt.

Fourteen (52%) of the 27 respondents answered "yes", 3 (11%) answered "no" and 10 (37%) had no opinion on the subject (Table 102). The majority of the responses contained information about how the bears now occur closer to the coast. Half of these respondents were of the opinion that the change to the distribution is due to changing ice conditions. However, some of the answers indicated that there has been an increase in number of bears that are caught from skiffs (Table 103).

The majority of the responses expressed the idea that the bear hunt has changed because the polar bears occur closer to inhabited areas and areas that are commonly used for hunting. A 27-year-old hunter from Innaarsuit put it as follows:

> I think that is the reason why more bears have been caught ... for example, in the spring in the past hunters could travel a long way to the west – when I lived in Upernavik Kujalleq/Søndre Upernavik we could travel a long way to the west without seeing bear tracks. Now lots of bears are caught close to the land in the spring. Just at the edge of the shore. [........] Back then, the bears were rare in Upernavik Kujalleq/Søndre Upernavik. I can tell you that for certain. During my childhood they were very rare. Then around the time I moved here, in about 1999, the hunters started to hunt a lot of bears.

	Total	Nutaar-miut	Innaar-suit	Naajaat, Tussaaq Aappilattoq	Upernavik town
Yes	14	1	4	5	4
No/Same	3	1	1	1	-
No opinion	10	1	3	4	2
N_{total} (hunters asked)	27	3	8	10	6

Table 102. Distribution of responses to question 25, from the area to the south of Nuussuaq (Area 4) in the Upernavik municipality: "Have the changes affected the polar bear hunt?"

	Number of responses	Nutaar-miut	Innaar-suit	Naajaat, Tussaaq Aappilattoq	Upernavik town
Closer	4	-	2	1	1
- because of changes to the ice	4	-	1	1	2
Changes to the methods	1	-	-	1	-
Others	4	-	1	2	1
N_{total} (responses)	13	0	4	5	4

Table 103. Types of responses to question 25, from the area to the south of Nuussuaq (Area 4) in the Upernavik municipality: "Have the changes affected the polar bear hunt?"

A 42-year-old hunter from the same settlement stated that polar bears can be encountered even when hunters are not out to hunt bears, and a 46-year-old from Aappilattoq was of the opinion that more bears are "on the way towards the land".

The fact that the hunting patterns have changed because of changes to the ice coverage was mentioned by several respondents. A 54-year-old from Nutaarmiut and a 44-year-old from Innaarsuit both stated that it is difficult to travel west by dogsled on hunting trips in spring because of the leads in the ice. One of these informants added that the cracks have become more problematic as a result of stronger currents. A 50-year-old from Aappilattoq believed that the decrease in sea ice has partly caused a change in the distribution of polar bears and has also meant an increase in hunting from skiffs:

> [........] Because the sea doesn't really freeze over, so the bears come closer. In the past, there used to be a lot of ice in this area until June and the bears were a long way from here. They were to the west of Upernavik. They seldom came by here. [........] The fact that the weather has got warmer had meant that the skiffs ... they get to the leads out there and sail on the open water. In the leads, there are narwhals and ... there are lots of game animals. Loads and loads of seals, loads of birds ... out there, and lots of bears. It seems like there are lots of bears.

An 18-year-old hunter from Upernavik also stated that dogsled hunting on the drift ice had become less common because of the poor ice conditions.

Changes to hunting practices were also mentioned by other informants. A 40-year-old informant from Aappilattoq stated that more bears are now caught from boats because the sea ice melts earlier. A 49-year-old hunter from the town of

Upernavik commented upon bear hunting from a sled as opposed to a boat:

> Yes. The changes have had consequences. For example, we never go bear hunting in that area anymore. [........] When we are doing something else, then the bears come and disturb us – *pakasaallutik*. I was once visited by a bear while I was in a tent. So I had to shoot at it. [1998....] It is in that way that the bears sometimes disturb us – *pakasaasut*. And the bear hunt from dogsleds ... that is ... if there were to be limits on sled hunting, I wouldn't like it. They ought instead to limit the polar bear catch from boats [........]. It is only the real hunters that go bear hunting using sleds. It is very seldom that they catch more bears than the hunters that use boats. It is much more difficult to hunt from a sled. [........] Consider the bear hunt from skiffs in the spring. That is when most of them are caught, and any hunter can catch them in this way. So anyone with a skiff can catch a bear, as long as he is with someone who has a professional hunting licence. It is only the dogsled that... Most people go out to sea ... that is ... some bears are lost when they reach the leads. There are lots that save their life in that way. Only a minority are caught on the ice, when the ice is good. Therefore, the hunting from dogsleds ought not to be limited. After all, they are the real hunters. It is our clothes, for example, I go hunting on a dogsled with bearskin trousers – not this kind of clothing (points to the boiler suit he is wearing during the interview; author's note). Nowadays there are people with this kind of clothing who ... skiff hunters.

A couple of hunters mentioned the relationship between the increased occurrence of seals and the bear catch. A 68-year-old hunter from Innaarsuit explained how the polar bears arrive in large numbers in the autumn, when there are many ringed seals. He expressed this as follow:

> That was their sign (*i.e.* the ancestors; author's note) for when the bears would come in large numbers, that is, when there were many ringed seals.

A 48-year-old hunter from Upernavik noted that the seals now travel further to the east (which attracts bears; author's note), perhaps to look for food or maybe due to changes in weather and ice conditions. He added however that it may be the case that there are many more bears.

Photo: E.W. Born

Some of the other responses were less specific and merely hinted that more bears are caught. A 75-year-old hunter from Nutaarmiut did not answer the question directly, but indicated that fewer polar bears "visit" because of the noise from skiffs.

26. Have there been any changes to the polar bears that you have seen or caught?

Twenty-four (96%) of the 25 respondents answered "no" and 1 (4%) from Nutaarmiut had no opinion on this matter. Three hunters said that generally the bears are fat and one of them recalled that in two years (1998 and 2000), when there had been very little ice, only two of the bears that he had caught had been very thin. A 27-year-old hunter from Innaarsuit answered the question in the negative, but also added that the number of young bears (2-year-olds) has increased because there are now:

> So many bears. In the past, that is ... during my childhood, and perhaps before I was born, bears were seldom caught. And the tracks that were observed in the past were always tracks from large bears. Maybe because there were fewer bears.

27. Have you observed changes to the occurrence of polar bears?

Nineteen (70%) of the 27 respondents answered this question by saying "yes", 4 (15%) said "no" and the remaining 4 (15%) had no opinion on the subject (Table 104).

	Total	Nutaar-miut	Innaar-suit	Naajaat, Tussaaq Aappilattoq	Upernavik town
Yes	19	2	8	5	4
No / Same	4	-	-	3	1
No opinion	4	-	1	2	1
N_{total} (hunters asked)	27	2	9	10	6

Table 104. Distribution of responses to question 27, from the area to the south of Nuussuaq (Area 4) in the Upernavik municipality: "Have you observed changes to the occurrence of polar bears?"

	Number of responses	Nutaar-miut	Innaar-suit	Naajaat, Tussaaq Aappilattoq	Upernavik town
Closer	15	2	6	4	3
- because of changes to the ice	-	-	-	-	-
- because there are more of them	6	1	3	1	1
Other	-	-	-	-	-
N_{total} (responses)	21	3	9	5	4

Table 105. Types of responses to question 27, from the area to the south of Nuussuaq (Area 4) in the Upernavik municipality: "Have you observed changes to the occurrence of polar bears?"

The majority of the responses included statements indicating that the bears have come closer and are more often sighted in the "local" area, but around one fourth of the answers indicated that the number of bears has generally increased (Table 105).

Several respondents mentioned that the bears increasingly occur in coastal areas. A 45-year-old hunter from Nutaarmiut was of the opinion that the bears have come closer in the last 20 years and several hunters from Innaarsuit expressed the same opinion in their responses (Table 105). A number of informants elaborated on their response. A 68-year-old resident of Innaarsuit, who had been the chairman of the local section of the Organization of Fishermen and Hunters in Greenland (KNAPK) on a number of occasions and had therefore kept abreast of the catches, said:

> However, in the early 1980s or if I think about the early 1970s, bears were so rare in those years that perhaps four or five bears were caught here, at most. Back then bears used to be a rare catch. But in later years, from 1990 to 1995 there were markedly more bears in our coastal areas. That's for sure. So at that time we went hunting

> a lot … in the early 1980s … that is, around Kingittuarsuk and Kitsissut, the central areas for our hunting trips, you know. That is where we had our seal nets. Back then, bear tracks were very, very rare. Even if we travelled a long way to the west. Rare … But nowadays, when we attend to our nets in that area, there are so many stray bears, that in the course of just a few days, the bears leave many tracks there. It is even the case that one might say that it is dangerous to go on a day trip without a weapon nowadays. Bears can come unexpectedly.

A 66-year-old man from the same settlement was of the opinion that the reason for bears' proximity to the coast is that they flee the noise created by shipping traffic in the west, and therefore wander towards food near land. A 30-year-old hunter, also from Innaarsuit, said that "these days" a hunter can observe as many as six bears in just one day.

In Aappilattoq, a 35-year-old hunter had noticed that there have been a greater number of bears to the west, north and south of Upernavik town in recent years and added that this had not been the case 5–10 years ago. He also stated that hunters have started to catch a larger number of bears in May and June in the last five years (see Figs. 36, 37). A 40-year-old informant from the same place indicated that there is evidence of both an increase in hunting activity involving boats and an increase in the occurrence of polar bears:

> I have only noticed that things may not have changed that much, maybe. For example, by sailing more on the open sea, they have become easier to catch now. Now that we can travel anywhere at all, we see them more often now … it seems like we see them more often now … it seems like, you know … there are far more tracks along the Upernavik coast.

Among the hunters from Upernavik town who commented on the change to the occurrence of bears there was a 49-year-old man who explained:

> As soon as the ice forms, then we start to catch them close to here. And as soon as the ice melts, they start to catch lots of bears right here [........] The bears are so close now. For example, my father who was born in 1930 … he is from Kangersuatsiaq. The place where my father attends to his nets … he goes there on foot … here in Maniitsuarsuk. During the 1940s and 1950s there were no bears there. Nowadays …

> nowadays there are lots of bears in that area. Back then there were none. There were very rarely bear tracks there.

Some of the responses to this question were actually statements indicating that the number of bears has increased (Table 105). The 68-year-old hunter from Nutaarmiut stated that when he "had become aware of bears" in 1958, bears were not often heard of, but that since 1982 when he had caught his first bear, bear sightings became more frequent. He did however add:

> But on the other hand, one might say that the bears maybe come for the ice. Or perhaps they seek food there [........] I don't think that there are fewer of them.

A 53-year-old hunter from Tussaaq said:

> ... During the last 20 years, many bears have come to the east ... there are so many of them that I can tell you that sometimes, when I see people walking on the ice, I want to warn them that they ought to look out, you know. Bears can move from one place to another very quickly.

A 48-year-old resident of Upernavik town who was one of the hunters who thought that there had been no change to the occurrence did however indicate that the level of hunting effort can affect people's perception of how many bears there are:

> If you are a hunter and travel all over the place, then you may encounter bears everywhere [........]. Those who look for them also catch them more often. If you don't look for them, then you also find them less frequently.

The biology and behaviour of polar bears

28. Have you seen mating, mating behaviour or tracks that indicate these behaviours?

None of the 27 respondents to this question had actually seen polar bears mating, but four hunters (from Nutaarmiut, Innaarsuit, Tussaaq and Aappilattoq, respectively) had seen tracks which indicated mating behaviour or imparted some knowledge on the subject.

A hunter from Nutaarmiut and his hunting partner had shot two bears in mid–

Tracks from a mating pair of polar bears on the pack ice ca. 60 km west of Kullorsuaq, 10 April 2010. Photo: K. L. Laidre

April, which seemed to be a mating pair from their tracks. A 68-year-old hunter from Innaarsuit understood through conversations with other hunters that the male bears catch the scent of females with cubs and follow their tracks. This takes place in March because of the prevailing north wind. This behaviour is called *tussaattut*, *i.e.* that a male follows in her tracks (*tumit* = tracks, *tussaattut* = follow each others' tracks). A 53-year-old hunter from Tussaaq recounted how he had once come across the tracks of a large male that was following a female with 2-year-old cubs, apparently to chase the cubs away in order to mate with her a long way out on the drift ice. His view was that the bears mate on the dense drift ice in April. A hunter from Aappilattoq was able to recount that he had seen tracks of mating bears 2–3 times in late March and April.

29. What do the bears eat, apart from ringed seals?

Of the 28 hunters who were asked this question, 19 (68%) were able to talk about observations of other foods than ringed seals, whereas 9 (32%) had only seen ringed seal remains in bears' stomachs (Table 106).

Various vegetable foods had been observed (Table 107). A hunter from Upernavik said that the bears can eat "something from the ground" (moss?, grass?; author's note), while a hunter from Innaarsuit had seen bears digging for kelp out on the drift

	Total	Nutaar-miut	Innaar-suit	Naajaat, Tussaaq Aappilattoq	Upernavik town
Yes	19	2	5	6	6
No opinion	9	1	4	4	-
N_{total} (hunters asked)	28	3	9	10	6

Table 106. Distribution of responses to question 29, from the area to the south of Nuussuaq (Area 4) in the Upernavik municipality: "What do the bears eat, apart from ringed seals?"

ice. Seaweed was cited as being an alternate food by a couple of other informants.

Four informants stated that young bearded seals (*teqillik*) and walruses can also be prey for polar bears. One informant spoke of how he had found 30 cm long strips of bearded seal skin on the ice, which seemed as though they had been ripped up by bear claws. Narwhals are also taken by bears (Table 107). A 49-year-old hunter from Upernavik said that he had seen a polar bear hauling a narwhal up onto the ice. According to him a bear can easily take a "white" narwhal (*i.e.* older animals) and narwhals with a tusk (*i.e.* males) as well as individuals that weigh "several tons" (an adult male narwhal can weigh up to ca. 1.8 tons; Heide-Jørgensen & Laidre 2006). Another informant spoke of having found the remains of a young narwhal's skull, which had been left on the ice by a bear.

Two hunters talked about cases where bears had parts of other bears in their stomachs, and a third man spoke of how males can kill a mother bear in order to first eat her cubs and then mate with her.

Greenland halibut had also been observed in the stomachs of bears and in a couple of cases they had taken frozen fish from fishing lines. In one case from Upernavik town, a hunter described how a bear had eaten common eiders at a hole in the ice close to an iceberg.

Other food (apart from ringed seal)	Number of responses	Nutaar-miut	Innaar-suit	Naajaat, Tussaaq Aappilattoq	Upernavik town
Grass, plants, moss	1	-	-	1	-
Seaweed	3	-	1	1	1
Other seals and walruses	4	-	-	1	3
Whales	5	1	-	1	3
Bear cubs	3	-	2	1	-
Other (*e.g.* birds and fish)	5	1	1	2	1
N_{total} (responses)	21	2	4	7	8

Table 107. Types of responses to question 29, from the area to the south of Nuussuaq (Area 4) in the Upernavik municipality: "What do the bears eat, apart from ringed seals?"

An adult ringed seal that was killed by a polar bear. Polar bears prefer to eat the blubber and the skin off the seal. Hunters have noted that polar bears often cool their prey in the snow and ice before eating it. Photo: K. L. Laidre

30. Have you seen polar bears hunting?

Nine (33%) of the 27 hunters who were asked this question said "yes", 16 (60%) said "no" and 2 (7%) did not give a clear statement on the subject (Table 108).

Six of the nine affirmative responses consisted of observations of bears hunting ringed seals, either at their breathing holes (*kikkuleq*), in birth lairs or as *uuttoq* (Table 109). However, most of the observations were lacking in detail because they were made under circumstances where the bear had been disturbed because it was being hunted.

A hunter from Innaarsuit described how the bears sniff out breathing holes:

> Yes, I have seen a bear hunting. I secretly watched it. Then ... it was looking for breathing holes. It sought seals' breathing holes. When the bear wanders around oblivious to the fact that it is being observed, then it does nothing but sniff, that's all it does. When it

	Total	Nutaar-miut	Innaar-suit	Naajaat, Tussaaq Aappilattoq	Upernavik town
Yes	9	-	4	3	2
No	16	3	4	6	3
No opinion	2	-	-	1	1
N_{total} (hunters asked)	27	3	8	10	6

Table 108. Distribution of responses to question 30, from the area to the south of Nuussuaq (Area 4) in the Upernavik municipality: "Have you seen polar bears hunting?"

	Number of responses	Nutaar-miut	Innaar-suit	Naajaat, Tussaaq Aappilattoq	Upernavik town
Ringed seal - at breathing holes	4	-	3	1	-
- in their lairs	1	-	-	-	1
- as *uuttoq*	1	-	-	1	-
Other	3	-	1	1	1
N_{total} (responses)	9	0	4	3	2

Table 109. Types of responses to question 30, from the area to the south of Nuussuaq (Area 4) in the Upernavik municipality: "Have you seen polar bears hunting?"

> doesn't know that it is being observed, you know. It crouches ... changes directions ... on the sea ice or close to the sea ... it just sniffs. It does nothing else. But at that time, I don't know what it was; it seemed as though it wanted to find something ... or acted like it was targeting something. That is how it wanders around. As if it is seeking something, sniffing it out ... onwards. I don't see them as bears that are hunting. But the ones that hunt, they can also be recognised by their tracks. Everything, even the tiniest thing ... it sniffs it out.

An informant from the town of Upernavik talked about his observations of a bear that was hunting ringed seal pups:

> ... I have watched one. Those *aninerit* ... it was hunting *aninerit* [........] They scan the surface of the ice like that [........] On the ice ... Using only its nose ... only using its sense of smell. Those that we cannot detect, which were in the pack ice ... (seal pups; author's note). It hauled the pups up from the ice. Obviously the seal mother cannot do anything. It eats them. We call them *nunarsat* (seals' birth lairs under the snow; author's note), that is ... how ... they eat them, for example, *aninerit*. Bears eat lots of seals...

An informant from Aappilattoq described how he had seen a bear sneaking up on an *uuttoq*:

> It had its head down first, hid its cub, then down into position, it was far away. It caught the seal a long way out. When it was about to eat it, we caught it.

Among other brief observations of hunting behaviours, there was a case of a bear which had killed a ringed seal and then proceeded to play with it by throwing it high up in the air.

31. Have there been any changes to what the polar bears eat?

A total of 26 hunters were asked this question. Ten (38%) of the respondents answered "no" and the remaining 16 (62%) had no opinion on the subject. One hunter said that they eat everything.

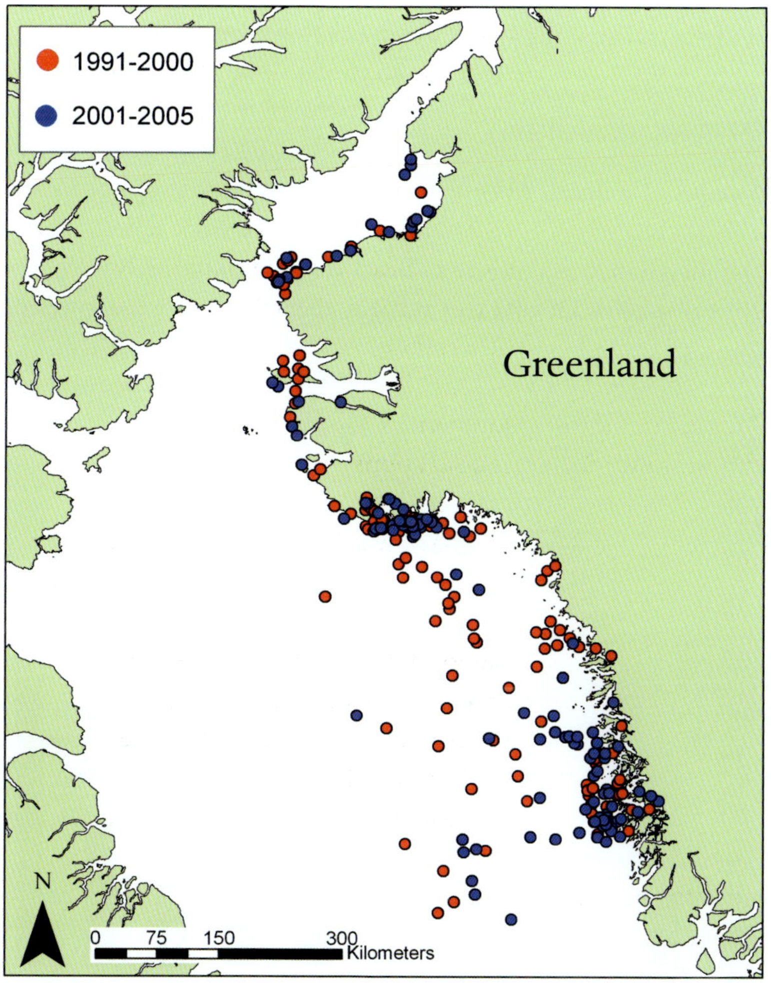

FIG. 29.
The distribution of 293 polar bear catches in the Qaanaaq and Upernavik municipalities shown for two periods (1991–2000: n = 145; 2001–2005: n = 148). Data from the interviews in February 2006.

The catch in figures

In this section we give an overview of the polar bear catch in the Qaanaaq and Upernavik municipalities based on the data from the individual bear catches that were reported by the informants. The purpose is to present information on the geographic and seasonal distribution of the catch as well as the age and sex composition.

During the interviews the informants reported a total of 588 different catches (754 individual bears) taken between 1952 and 2005. About 60% of the catches took place between 1991 and 2005 and ca. 35% took place between 2001 and 2005. The distribution of the catches for which there was information on location since 1990 is illustrated in Fig. 29. In the following sections, the catch in the two municipalities is described separately and from north to south by area (1–4).

Qaanaaq municipality

Number of animals taken

Detailed data was collected about 345 polar bear catches during the period between 1952 and 2006. During the individual catches (*i.e.* hunting trips or "cases") between one and seven bears were taken.

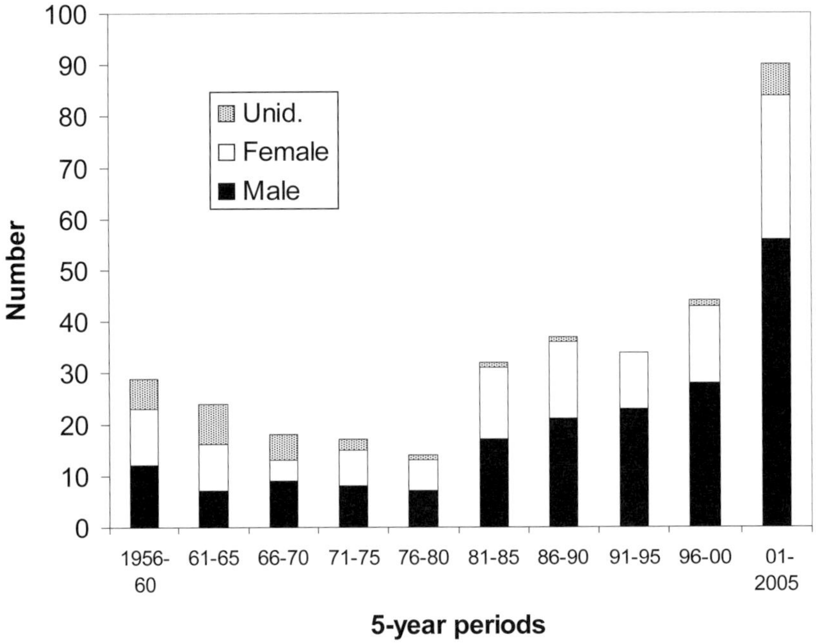

FIG. 30. Distribution by 5-year periods of the kills of 339 adult polar bears (329 catches) in the Qaanaaq municipality since 1952. Unid. = unidentified, *i.e.* sex was not specified; 1 adult male that was killed in 1952 is included in the period 1956–1960. Data from the interviews in 2006.

However, in 14 cases, the same bear catch had been reported by more than one hunter (for example, one catch of a female with two cubs was reported by four different hunters). These cases were omitted, which resulted in 331 polar bear catches (1952–2006). Furthermore, two catches from 2006 were excluded, which then comprised a total of 329 catches during which 415 polar bears (including cubs) were taken. Around 51% of these catches took place between 1991 and 2005 and ca. 27% (n = 90) took place between 2001 and 2005; Fig. 30.

During the interviews, a total catch of 14 to 36 animals per year (2001–2005) was reported. The highest catch was reported in 2005 when 14 informants reported 16 different polar bear catches. In every year, the total catch reported during the interviews was lower than the number stated in *"Piniarneq"* (Table 110).

Year	Number				
	Bears	Family groups[1)]	Cubs[2)]	Informants[3)]	*Piniarneq*[4)]
2001	18	3	6	6 (9)	20
2002	20[5)]	1	1	6 (7)	36
2003	14	1	1	10 (11)	47
2004	22	5	8	8 (12)	29
2005	36	4	5[6)]	14 (17)	50[7)]

1) Family groups: *i.e.* mothers with cubs; included in the total number of "bears"
2) Number of cubs in family groups
3) Number of interviewees (number of hunters who had killed bears)
4) The catch according to the *"Piniarneq"* catch reporting system (Department of Fisheries, Hunting and Agriculture, DFHA, Nuuk)
5) Includes 6 animals without any further information
6) 2 cubs in a family group were not included, as they got away during the hunt
7) 25 of these were reportedly caught in Savissivik and 25 in the area to the north (Source: DFHA; 7 January 2007)

Table 110. Information on the number of polar bears, family groups and cubs (>1 year old) killed between 2001–2005, obtained during the interviews in the Qaanaaq municipality in February 2006. The number of informants and the catch according to catch recording system *"Piniarneq"* are also shown.

Interviewees were asked to estimate the number of bears taken in 2005 by hunters from their settlement. Nineteen informants were able to give a more or less exact estimate of the catch: Siorapaluk: 6 to >10; Qaanaaq and Qeqertarsuaq: 7 to >10; Savissivik: 20 to 30 (probably closer to 30). These figures indicate that 35 to 50 polar bears were taken in the Qaanaaq municipality in 2005.

The geographic distribution of the catch

Ikersuaq/Kane Basin (Area 1) and Qimusseriarsuaq/Melville Bay (Area 3) have traditionally been the primary polar bear hunting areas (Vibe 1968, Rosing-Asvid & Born 1990). However, in recent years, the polar bear hunt in Qimusseriarsuaq seems to have increased in importance (Table 111).

Until 1967, when Greenlandic hunters were banned from hunting polar bears in Canadian territory (Kiliaan *et al.* 1978), the hunt in Area 1 was the most significant.

During the interviews, data pertaining to a total of 51 polar bear catches from Umimaat Nunaat/Ellesmere Island in Canada were gathered. Thirty-eight of the reported catches in this area took place between 1959 and 1967, and the remainder occurred between 1968 and 1985 (the last three catches took place in 1985). However, it must be emphasised that in several cases the information about the exact location of the catch on or around Umimaat Nunaat/Ellesmere Island was unclear.

Qimusseriarsuaq/Melville Bay has been the principal hunting area since approximately 1990. The catch in the central area (Area 2) is characterised by being more incidental, as hunters tend to kill bears that they come across when hunting other species. However, since the 1980s, the percentage of bears that are taken in Area 2 appears to have increased (Table 111).

Our survey shows that hunters from Siorapaluk primarily travel to the north in order to hunt bears in the northern reaches of Ikeq/Smith Sound and in Ikersuaq/Kane Basin, but that they also hunt polar bears to some extent in Area 2 (Table 112). The hunters who live in Qaanaaq and Qeqertarsuaq/Herbert Island hunt polar bears

10-year and 5-year period	Area				Area		
	1	2	3	Total	1	2	3
	n	n	n	n	%	%	%
1952–1960	19	0	10	29	65.5	0.0	34.5
1961–1970	31	1	10	42	73.8	2.4	23.8
1971–1980	12	2	17	31	38.7	6.5	54.8
1981–1990	25	19	25	69	36.2	27.5	36.2
All (1952–1990)	87	22	62	171	50.9	12.9	36.3
1991–1995	8	5	21	34	23.5	14.7	61.8
1996–2000	8	8	28	44	18.2	18.2	63.6
2001–2005	36	14	40	90	40.0	15.6	44.4
All (1991–2005)	52	27	89	168	31.0	16.1	53.0

Table 111. Distribution according to area (1–3) of 339 polar bears killed during 329 hunts in the Qaanaaq municipality presented in 10- and 5-year periods. Data collected during interviews in the municipality in February 2006.

in all of the sub-areas in the municipality. It is however striking that the percentage of bears that they have killed in Area 2 ("the local area") has increased markedly over the last 15 years (1952–1990: 18.7%, $n_{total} = 76$; 1991–2005: 58.5%, $n_{total} = 41$). Hunters from Savissivik reported bear catches in Area 3 only (Table 112).

Period	Area	Siorapaluk		Qaa.-Qeq.		Savissivik		Kull.-Nuus.		S. of Nuus.	
		Number	%	Number	%	Number	%	Number	%	Number	%
1952–1990	1	38	16.6	45	19.7	0	0.0	0	0.0	0	0.0
	2	8	3.5	14	6.1	0	0.0	0	0.0	0	0.0
	3	1	0.4	16	7.0	45	19.7	26	11.4	4	1.7
	4	0	0.0	0	0.0	0	0.0	2	0.9	30	13.1
1991–2005	1	34	9.6	12	3.4	0	0.0	0	0.0	0	0.0
	2	3	0.8	24	6.8	0	0.0	0	0.0	0	0.0
	3	1	0.3	5	1.4	83	23.5	54	15.3	7	2.0
	4	0	0.0	0	0.0	0	0.0	0	0.0	130	36.8

Table 112. Polar bear catches (n = 582), for which the year was stated, given by sub-area (1–4) and towns/settlements in the Qaanaaq and Upernavik municipalities and shown for two periods 1952–1990 and 1991–2005. Data collected during interviews in the municipality in February 2006.

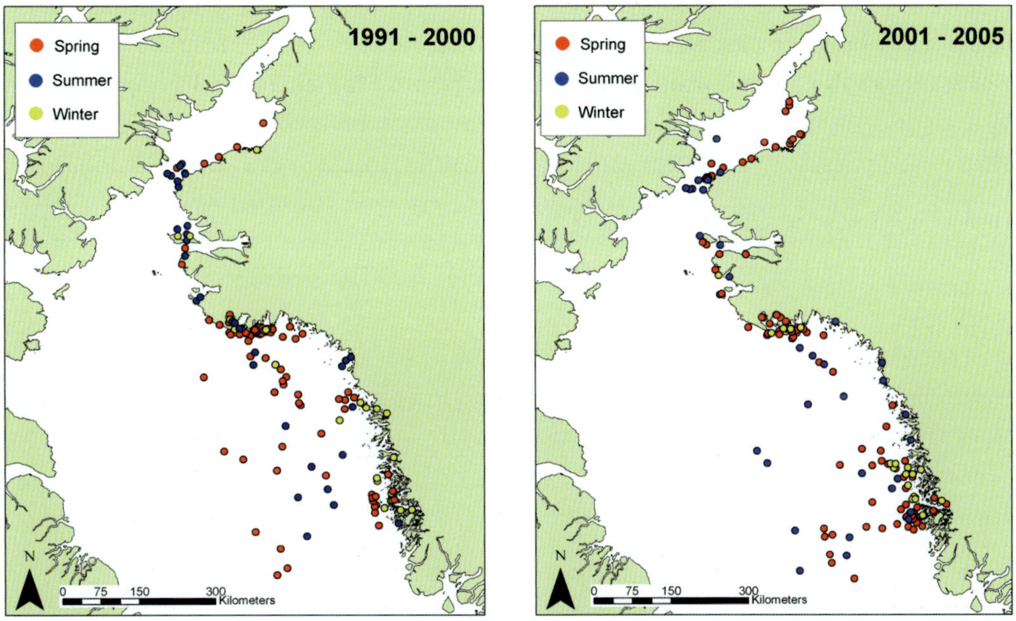

FIG. 31. Distribution of 244 polar bear catches in the Qaanaaq and Upernavik municipalities according to season and period (see: "Materials and methods"; 1991–2000: n = 115; 2001–2005: n = 129. Data from the interviews in 2006.

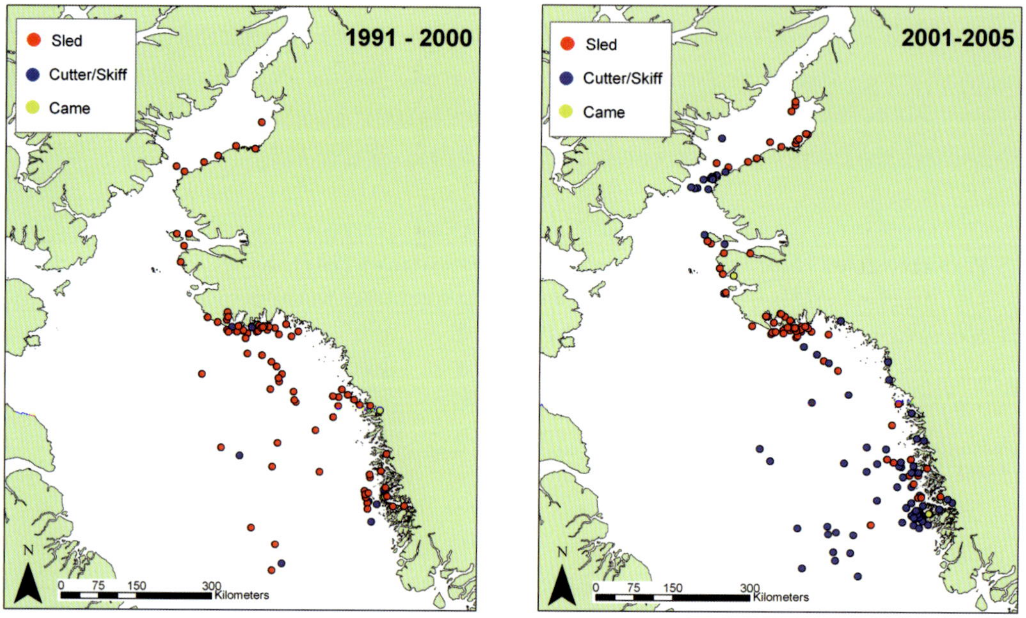

FIG. 32. Distribution of 215 polar bear catches in the Qaanaaq and Upernavik municipalities according to hunting method and period (1991–2000: sled, cutter and arrived of own accord, respectively, n=76, 7, 1; 2001–2005: 61, 68 and 2. Data from the interviews in 2006.

Compared to the period 1991–2000 there has not been any conspicuous change in the geographic distribution of the catch between 2001 and 2005 and the seasonal distribution of bears caught has also remained unchanged (Figs. 29, 31). In this municipality there has seemingly been no change to the areas in which bears are killed from dogsleds and the areas where they are shot from skiffs (Fig. 32). There has, on the other hand, been a change to the times of year when the two modes of transport are employed in hunting activities (see the section entitled "Hunting methods", p. 187).

There were two reported cases (2000, 2005) of polar bear catches in Melville Bay Nature Reserve Protection Zone II, where hunting is forbidden. Both catches took place from skiffs in July and August, respectively.

Seasonal distribution

In the period between 1952 and 1990, around 75% of the catches in the Qaanaaq municipality took place between February and April, with a peak in the month of April. Around 21% of the catches occurred from October to November (Fig. 33). After 1990, this pattern changed. Between 1991 and 2000, the catch of polar bears

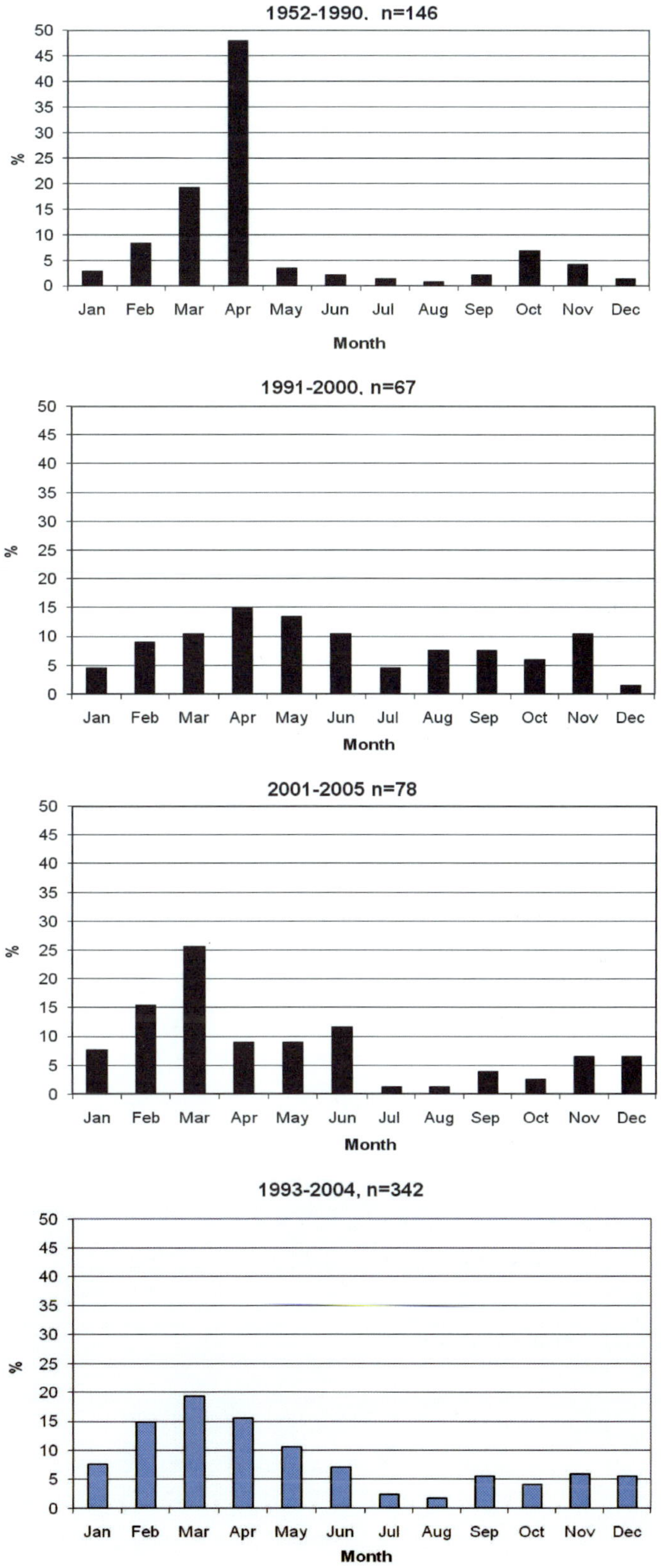

FIG. 33. The polar bear catch by month in the Qaanaaq municipality (three time periods in 1952–2005), as specified in the interviews in February 2006 (three upper panels) together with the catch recording system *"Piniarneq"* (1993–2004), bottom panel.

was spread over a larger part of the year, with around 58% of the catches made between February and June. Generally, this distribution also characterised the catch between 2001 and 2005, but in this period the catch took place somewhat earlier in the season, reaching a conspicuous peak in February/March (around 41% of the catch); Fig. 33.

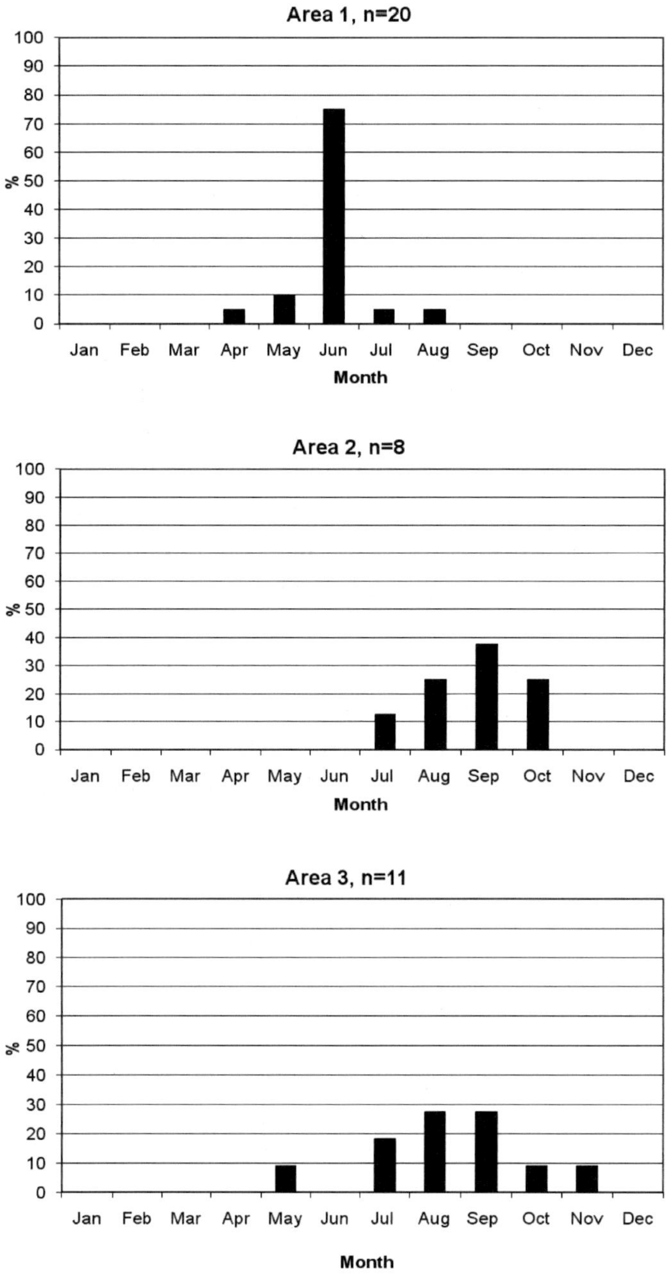

FIG. 34.
Seasonal distribution of polar bear catches from boats (1991–2005) shown for three areas (1–3) in the Qaanaaq municipality. Data from the interviews in 2006. During the interviews, six catches from boats prior to 1991 were also reported.

The trend towards a wider seasonal distribution of the catch in the last 15 years has in particular been a characteristic feature of the catch in Areas 1 and 3 (data not shown). In Area 1, the spring dogsled hunt tended to reach its peak earlier (in March instead of April), and there was a marked increase in hunting activities in June. The latter is due to the fact that a number of polar bears are taken during hunting trips by boat.

Hunting methods

Until 1990, around 92% of the bear catches in the Qaanaaq municipality took place from dogsleds. From the early 1990, however, this figure dropped to around 71% due to the concurrent increase in the proportion of polar bears caught from skiffs. It is especially in Area 1 where a marked increase in the percentage of bears caught from skiffs or cutters has been seen since 1990, but this trend has also been observed in Area 3.

In May and in particular June the hunters take their skiffs to the ice edge in Iluleerloq/Murchison Sound and from there they sail in a northerly direction to the southern edge of the pack ice in Ikersuaq/Kane Basin, which lies between Anoritooq/Kap Inglefield and Pikiuliusarsuaq/Pim Island in Canada. In recent years, polar bears have been shot at the ice edge during these hunting trips, during which walrus and narwhal and other species are hunted. In areas 2 and 3 the polar bear hunt from boats takes place in the summer and autumn in particular (Fig. 34).

Although the number of bears that have come into settlements and populated areas on their own accord has been very small the interview survey indicated a slight increase which seems to be most pronounced in Savissivik (Table 113). In three of the five reported cases, the bears appeared in the period between December and

Area	Period	Sled		Skiff/Cutter		Visiting		Unid.		Total
		n	%	n	%	n	%			n
1	1952–1990	87	97.8	2	2.2	0	0.0	0	0.0	89
	1991–2005	29	55.8	23	44.2	0	0.0	0	0.0	52
2	1952–1990	14	63.6	5	22.7	0	0.0	3	13.6	22
	1991–2005	17	63.0	9	33.3	1	3.7	0	0.0	27
3	1952–1990	59	95.2	2	3.2	1	1.6	0	0.0	62
	1991–2005	73	82.0	11	12.4	3	3.4	2	2.2	89

Table 113. Methods employed when hunting polar bears (n = 341) in three areas in the Qaanaaq municipality during two periods 1952–1990 and 1991–2005. Data collected during interviews in the municipality in February 2006.

March, and a very thin female bear with cubs showed up at a meat rack in Moriusaq in September (there was one case where the time of year was not specified).

Age and sex distribution

The percentage of adult female bears in the catch is of significance when determining the sustainable take from a population of polar bears (Taylor *et al.* 1987, 1988). Of the 308 "adult" or "independent" (for definitions of "adult" and "independent" see p. 20) bears (for which the sex is known) that were taken in the Qaanaaq municipality (1952–2005), ca. 38% were females and 62% males.

If the earlier period (1952–1990) is compared to the later period (1991–2005), there was no statistically significant difference in the distribution of sex in the catch in any of the sub-areas (χ^2-tests; $p > 0.133$); Table 114. The sex composition of the catch did however differ among the different areas. For the period 1991–2005 there was no statistically significant difference between Areas 1 and 2 in terms of the sex ratios of the catch ($\chi^2 = 0.167$; $P = 0.684$, $df = 1$). In these two areas female bears constituted 23% of the catch of adult animals and males constituted 77% ($n = 73$). However, in the same period, there were proportionally more adult female bears in the catch in Area 3 (females: 42%; males: 58%; $n = 88$). This distribution is significantly different to the distribution in Areas 1 and 2 combined ($\chi^2 = 4.924$; $P = 0.025$, $df = 1$) and may indicate that Qimusseriarsuaq/Melville Bay is a denning

Driving by dogsled to the edge of the land fast ice at Appat/Wolstenholme Island, May 1991. Photo: E.W. Born

Table 114. Distribution of independent or "adult" polar bears according to location and sex (n = 341; 308 of these with information of sex) caught during 329 catches in the Qaanaaq municipality shown in 10- and 5-year periods (1952–2005). Data collected during interviews in the municipality in February 2006.

10- or 5-year period	Area 1						
	F n	M n	Unid. n	Total	F %	M %	Unid. %
1952–1960	7	6	6	19	36.8	31.6	31.6
1961–1970	7	13	11	31	22.6	41.9	35.5
1971–1980	5	7	0	12	41.7	58.3	0.0
1981–1990	7	17	3	27	25.9	63.0	11.1
All (1952–1990)	26	43	20	89	29.2	48.3	22.5
1991–1995	1	7	0	8	12.5	87.5	0.0
1996–2000	3	5	0	8	37.5	62.5	0.0
2001–2005	6	24	6	36	16.7	66.7	16.7
All (1991–2005)	10	36	6	52	19.2	69.2	11.5

10- or 5-year period	Area 2						
	F n	M n	Unid. n	Total	F %	M %	Unid. %
1952–1960	0	0	0	0	0.0	0.0	0.0
1961–1970	0	0	1	1	0.0	0.0	0.0
1971–1980	1	0	1	2	50.0	0.0	50.0
1981–1990	7	12	0	19	36.8	63.2	0.0
All (1952–1990)	8	12	2	22	36.4	54.5	9.1
1991–1995	2	3	0	5	40.0	60.0	0.0
1996–2000	2	6	0	8	25.0	75.0	0.0
2001–2005	3	11	0	14	21.4	78.6	0.0
All (1991–2005)	7	20	0	27	25.9	74.1	0.0

10- or 5-year period	Area 3						
	F n	M n	Unid. n	Total	F %	M %	Unid. %
1952–1960	4	6	0	10	40.0	60.0	0.0
1961–1970	6	3	1	10	60.0	30.0	10.0
1971–1980	7	8	2	17	41.2	47.1	11.8
1981–1990	15	9	1	25	60.0	36.0	4.0
All (1952–1990)	32	26	4	62	51.6	41.9	6.5
1991–1995	8	13	0	21	38.1	61.9	0.0
1996–2000	10	17	1	28	35.7	60.7	3.6
2001–2005	19	21	0	40	47.5	52.5	0.0
All (1991–2005)	37	51	1	89	41.6	57.3	1.1

Area		Age of litter (years)[1)]			Adult females	Litters per adult female	Total catch
Municipality Qaanaaq		1	2-3	?	n		n
1	Number of litters (total number of cubs)	0	2(4)	0	8	0.25	46
	Number of cubs / litter: Mean / SD / min.-max.	0	2.0 / 1.41 / 1-3	0			
	Sex: Male / Female / ?	0	- / - / 3	0			
2	Number of litters (total number of cubs)	0	5(7)	0	7	0.71	27
	Number of cubs / litter: Mean / SD / min.-max.	0	1.67 / 0.55 / 1-2	0			
	Sex: Male / Female / ?	0	4 / 3 / -	0			
3	Number of litters (total number of cubs)	0	18(27)	2(3)	28	0.71	89
	Number of cubs / litter: Mean / SD / min.-max.	0	1.50 / 0.51 / 1-2	1.50 / 0.71 / 1-2			
	Sex: Male / Female / ?	0	14 / 5 / 8	- / - / 3			
1–3	Number of litters (total number of cubs)	0	25(38)	2(3)	43	0.63	162
	Number of cubs / litter: Mean / SD / min.-max.	0	1.52 / 0.58 / 1-3	1.50 / 0.71 / 1-2			
	Sex: Male / Female / ?	0	18 / 8 / 11	- / - / 3			
Municipality Upernavik							
3	Number of litters (total number of cubs)	3(5)	3(6)	2(3)	23	0.35	53
	Number of cubs / litter: Mean / SD / min.-max.	1.67 / 0.58 / 1-2	2.00 / 0.0 / 2	1.50 / 0.71 / 1-2			
	Sex: Male / Female / ?	1 / - / 4	- / - / 6	- / - / 3			
4	Number of litters (total number of cubs)	4(7)	14(25)	3(6)	48	0.44	129
	Number of cubs / litter: Mean / SD / min.-max.	1.75 / 0.50 / 1-2	1.79 / 0.58 / 1-3	2.00 / 0.0 / 2-2			
	Sex: Male / Female / ?	- / - / 7	5 / 2 / 18	2 / 2 / 2			
3+4	Number of litters (total number of cubs)	7(12)	17(31)	5(9)	71	0.41	182
	Number of cubs / litter: Mean / SD / min.-max.	1.71 / 0.49 / 1-2	1.82 / 0.53 / 1-3	1.80 / 0.45 / 1-2			
	Sex: Male / Female / ?	1 / - / 11	5 / 2 / 24	2 / 2 / 5			

1) No reported catch of 0-year olds

Table 115. Number of litters of different ages, average litter size and distribution according to sex in family groups of polar bears in the Qaanaaq and Upernavik municipalities (1991–2005). Data collected during interviews in the municipality in February 2006.

Municipality	Area	Adult				Adult	Old				Old	Young				Young	Total
		n	F	M	Unid.	%	n	F	M	Unid.	%	n	F	M	Unid.	%	
Qaanaaq	1	81	29	52	0	69.8	14	1	13	0	12.1	21	4	14	3	18.1	116
	2	26	11	15	0	55.3	2	1	1	0	4.3	19	3	16	0	40.4	47
	3	84	47	34	3	57.5	28	7	21	0	19.2	34	13	21	0	23.3	146
Upernavik	3	64	23	40	1	72.7	8	0	7	1	9.1	16	4	10	2	18.2	88
	4	106	47	59	0	63.1	16	1	15	0	9.5	46	5	40	1	27.4	168

Table 116. Age classification of 309 "independent" or adult polar bears (1952–2005) in the Qaanaaq municipality and of 256 adult polar bears in the Upernavik municipality (1957–2005). Data collected during interviews in the municipality in February 2006.

area. This theory is supported to some extent by a proportionally large number of females with cubs in the catch in this area (Table 115).

Data about the age category of independent animals (*i.e.* "adult", "old" and "young") showed no significant difference in the age distribution of the catch, in the periods 1952–1990, 1991–2000, and 2001–2005 (data are not shown) ($\chi^2 = 4.672$; $P = 0.335$, df = 4). However, the age distribution did vary in the different areas (Table 116; $\chi^2 = 15.143$; $P = 0.004$, df = 4). The age distribution did not differ significantly between Areas 1 and 3 ($\chi^2 = 4.417$; $P = 0.101$, df = 2). However, the age distribution in Area 2 differed significantly from the other two areas, as there were proportionally more "young" animals and fewer "old" animals in the catch in Area 2 ($\chi^2 = 13.903$; $P = 0.001$, df = 2; Table 116). This may reflect the fact that the bears taken in Area 2 are primarily stragglers, as younger bears generally tend to disperse and move over a larger area than older animals (*e.g.* Taylor *et al.* 2001). The relatively large percentage of "old" bears in the catch in Areas 1 and 3 may indicate fidelity of older polar bears to good feeding and denning habitat, as observed in other parts of the Arctic (*e.g.* Ramsay & Stirling 1990, Amstrup & Gardner 1994, Ferguson *et al.* 1997).

In some cases the hunters estimated the exact age of the bear that was killed. According to these estimates, "young" bears had an average age of 3.6 years (sd = 1.9; min.-max.: 2–9 years; n = 11), "adult" bears had an average age of 7.4 years (sd = 3.4; min.-max.: 3–15 years, n = 34), and "old" bears had an average age of 13.0 years (sd = 4.3, min.-max.: 8–16 years, n = 3).

Size and composition of litters

When there were reports of a catch involving a family group (*i.e.* a female bear with a cub / cubs), the hunters were asked to estimate the age of the cubs and to specify their sex. There were 27 reported family groups with a total of 35 cubs (1991–2005).

A young female polar bear being butchered on an ice floe near Pitoraarfik in June 1991. Photo: E.W. Born

The number of females with cubs varied a great deal from area to area, but the largest numbers were caught in Qaanaaq's Area 3 (Table 115) and the smallest in Area 1. This same trend also applies if the whole period from 1952 to 2005 is taken into consideration (data not shown).

In all of the cases (n = 25 litters) where the hunters specified the age of the cubs, the cubs were estimated to be around two years old (two litters were estimated to consist of three-year-old cubs) (Table 115). Of the 26 cubs (2-year-olds) for which the sex was specified, 69.2% were males and 30.8% were females.

However, in six cases prior to 1991, there were reports of females with 0-year-old cubs (4 cases) and 1-year-old cubs (2 cases) killed. In four of the cases, this had taken place prior to the protection of such family groups on 1 January 1975. After 1974, there was one reported catch in Jones Sound (Umimaat Nunaat/Ellesmere Island in Canada) in 1975 of a female with two cubs of around 1 year of age, where the dogs also killed the cubs. In another case from 1986, a female with two 0-year-old cubs was killed in Qimusseriarsuaq/Melville Bay by accident because of the poor light at dusk.

Upernavik municipality

Number of animals taken

In the Upernavik municipality, detailed information was collected for about 275 polar bear catches made between 1957 and 2006. During the individual catches, between one and four bears were taken. Six catches from 2006 (comprising a total of six adult bears) were excluded, as were three cases where hunters separately reported the catch of a cub which was part of a family group.

In seven cases, the same bear catch was reported by more than one hunter. These seven cases have also been extracted from the data, which therefore results in 259 catches (1957–2005) during which 339 bears (adults and cubs) were taken.

Around 75% of the reported catches occurred in the period from 1991 to 2005, and ca. 46% (n = 111) were reported during 2001–2005; Fig. 35. During the interviews, information was gathered on the catch of between 10 and 57 animals per year in the last 5 years. The highest yield was reported in 2005 (Table 117).

Twenty-one hunters were asked to estimate how many polar bears had been taken by hunters from their settlement in 2005 and 16 of them gave a more or less precise estimate which indicated the following yield sizes: Kullorsuaq: at least 20; Nuussuaq: at least 10; Nutaarmiut: 8; Innaarsuit: at least 10–40; Naajaat: 10–15; Tussaaq: "no

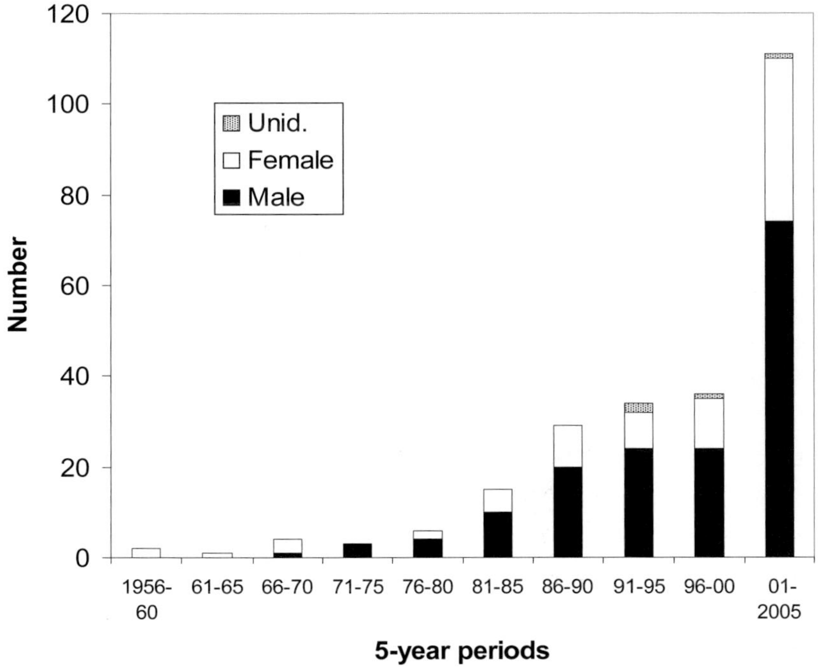

FIG. 35. Distribution by 5-year periods of the kills of 241 adult polar bears (241 catches) in the Upernavik municipality since 1957 (for an additional 13 catches, the year was unspecified and a further six catches took place in 2006) Unid. = unidentified, *i.e.* the sex was unspecified.

Year	Number				
	Bears	Family groups[1)]	Cubs[2)]	Informants[3)]	*Piniarneq*[4)]
2001	10	1	2	6 (7)	64
2002	15	2	4	7 (7)	73
2003	37	3	6	12 (12)	135
2004	24	6	10	10 (10)	85
2005	57	5	9	19 (19)	87

1) Family groups: *i.e.* mothers with cubs; included in the total number of "bears"
2) Number of cubs in family groups
3) Number of interviewees (number of hunters who had killed bears)
4) The catch according to the *"Piniarneq"* catch reporting system (source: DFHA, Nuuk, 7 Jan. 2007)

Table 117. Information on the number of bears, family groups and cubs (>1 year old) killed in the period between 2001 and 2005, obtained during interviews in the Upernavik municipality in February 2006. The number of informants and the catch according to the catch recording system *"Piniarneq"* are also shown.

10- or 5-year period	Area 3					Area 4				
	F n	M n	Total	F %	M %	F n	M n	Total	F %	M %
1952–1960	1	0	1	100.0	0.0	1	0	0	0.0	0.0
1961–1970	2	1	3	66.7	33.3	2	0	2	0.0	0.0
1971–1980	1	6	7	14.3	85.7	1	1	2	50.0	2.0
1981–1990	6	11	17	35.3	64.7	8	19	27	29.6	70.4
All (1952–1990)	10	18	28	35.7	64.3	12	20	32	37.5	62.5
1991–1995	3	9	12	25.0	75.0	5	15	20	25.0	75.0
1996–2000	4	10	14	28.6	71.4	11	24	35	31.4	68.6
2001–2005	9	16	25	36.0	64.0	36	93	129	27.9	72.1
All (1991–2005)	16	35	51	31.4	68.6	52	132	184	28.3	71.7

Table 118. Distribution according to area (3 and 4) and sex of "independent" or adult polar bears (n = 295) killed in the Upernavik municipality, shown in 10- and 5-year periods (1957–2005). Data collected during interviews in the municipality in February 2006.

estimate"; Aappilattoq: 10–15; Upernavik: ca. 10. These estimates indicate that in 2005, at least 80 polar bears were taken in the Upernavik municipality.

The geographic distribution of the catch

The percentage of the total catch, reported in the two sub-areas (3 and 4) in the Upernavik municipality, between 2001 and 2005 was highest in Area 4 (83.8% vs. 16.2% in Area 3). The catch in Area 4 was also proportionally higher in earlier times, but this tendency has been on the increase since around 1990 (Table 118).

Compared to the period 1991–2000, there has not been any conspicuous change to the geographic distribution of the catch in the Upernavik municipality, nor has

FIG. 36.
The polar bear catch by month in the Upernavik municipality (3 time periods in the period 1957–2005) as specified in the interviews in February 2006 (3 upper panels) as well as in the catch recording system *"Piniarneq"* (1993–2004), bottom panel.

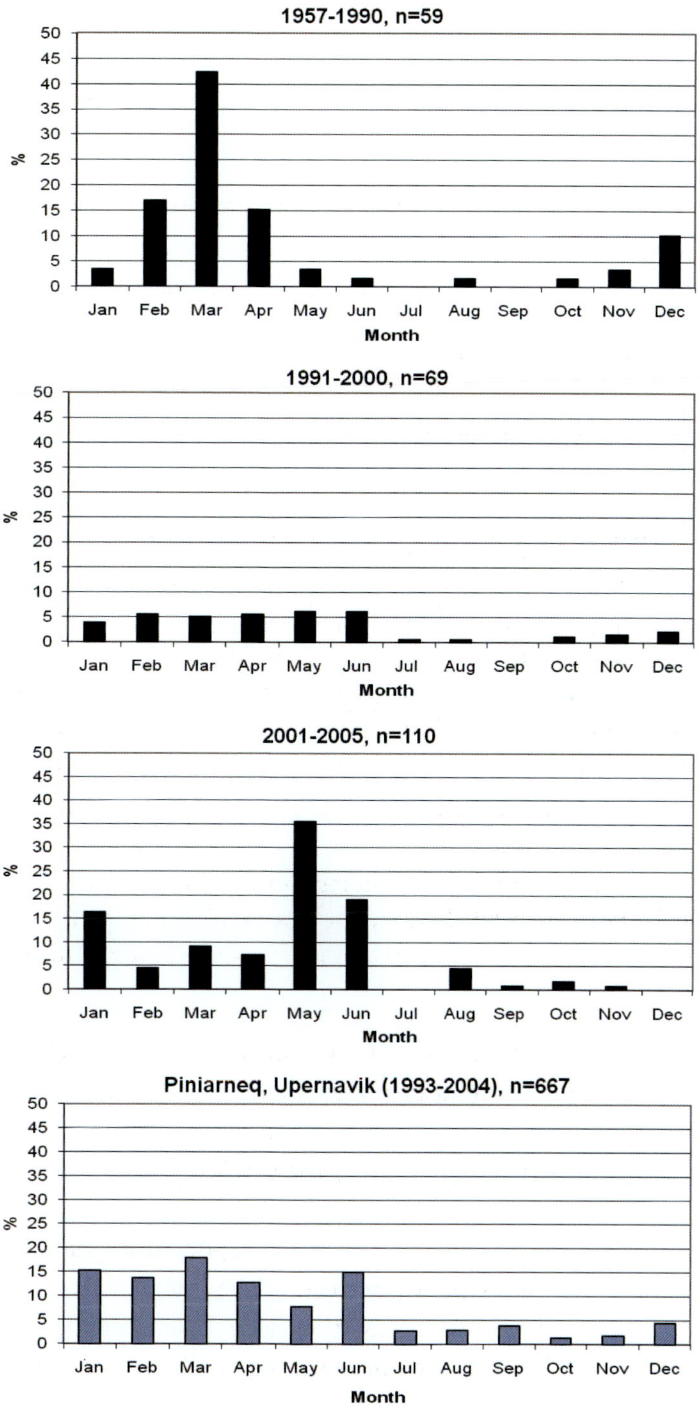

there been any change to their seasonal distribution (Fig. 32). However, there has clearly been a change to the methods that are employed in bear hunting in Area 4 (see the section on "Hunting methods" below).

There were reports of one case of polar bear hunting with a dogsled in April, inside the Melville Bay Nature Reserve Protection Zone II, where hunting is prohibited. The year was not specified, but judging from the age of the hunter, this probably occurred after 1990.

Seasonal distribution

In the period between 1957 and 1990, about 75% of the catches (44/59) in the Upernavik municipality took place during February–April, with a clear peak in the month of March. However, the pattern has been different since ca. 1990. During 1991–2005, 89% of the catch (159/179) was made in the period from January to June, with a peak occurring in May/June (around 46% of the catch). This peak in May–June is largely due to polar bear hunting from boats. The remainder of the catch was evenly distributed throughout the rest of the year (Fig. 36). The trend in the last ca. 15 years towards the catch being spread out over a larger part of the year was most pronounced in Area 3.

Hunting methods

In the period 1957–1990, about 92% of the bear catches in the Upernavik municipality took place using dogsleds, but in the subsequent period the percentage of the catch made by using boats increased significantly. During 1991–2005, ca. 38% of the catch in Area 3 and 63% of the catch in Area 4 involved the use of boats (Table 119). During 2001–2005 the percentage of bears that were caught from boats was 62% in Area 3 and 79% in Area 4. In Area 3, hunting from boats had mostly taken place in June and August, whereas in Area 4 this tended to occur in May and June (Fig. 37).

In Area 4, to the south of Nuussuaq, a number of the catches occurred further west than 60° W (more than circa 100 km from the coast). These catches were reported by hunters from Upernavik, Aappilattoq, Tussaaq and Innaarsuit. These "offshore" catches in the two periods (before and after 1990) also indicate an increase in hunting activity involving boats. Nine "offshore" catches prior to 1990 were all conducted using dogsleds in March and April. After 1990, 11 (30%) of the total of 37 catches that took place far out to sea involved the use of dogsleds, whereas the remainder involved the use of boats (ca. 77% of these catches that took place using boats occurred within the last 5 years). Since 1990, the catches involving dogsleds occurred between February and May (one in May), whereas the catches involving

Area	Period	Sled		Skiff/Cutter		Visiting		Unid.		Total
		n	%	n	%	n	%			n
3	1957–1990	28	93.3	0	0.0	2	6.7	0	0.0	30
	1991–2005	32	60.4	20	37.7	1	1.9	0	0.0	53
4	1957–1990	29	90.6	2	6.3	1	3.1	0	0.0	32
	1991–2005	38	29.2	82	63.1	5	3.8	5	3.8	130

Table 119. Methods employed when hunting "independent" or adult polar bears (n = 245) in two areas (3 and 4) in the Upernavik municipality during two periods (1957–1990 and 1991–2005). Data collected during interviews in the municipality in February 2006.

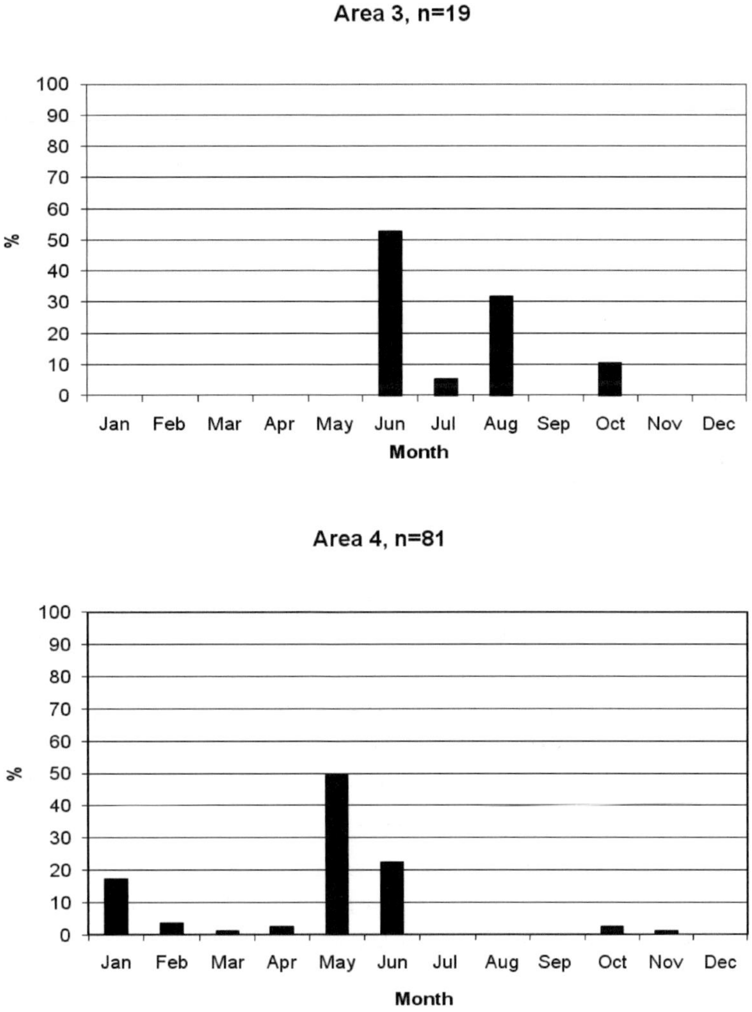

FIG. 37. Seasonal locations of polar bear catches from boats (1991–2005) shown for two areas (3 and 4) in the Upernavik municipality. Data from the interviews in 2006.

Skins of ringed seals being dried in the town of Upernavik. The catch of ringed seals is still of great importance in the Upernavik area. Photo: K. L. Laidre

boats all occurred in May (n = 15) and June (n = 11). Thus, the trend has been most pronounced during the last 5 years (Fig. 32). Poorer (less safe) ice conditions have made it difficult to travel by dogsled a long way offshore, whereas "lighter" ice conditions (more loose ice and open water) during the spring have made it possible to go polar bear hunting by boat relatively far out to sea, early in the season.

Of the 10 bears that came into inhabited areas, eight arrived in the period between November and January (one in September and one unspecified). If the phenomenon is considered over a longer period of time, there does not appear to be any trend to the prevalence of visits (Table 119).

Age and sex distribution

In the Upernavik municipality the sex distribution of the 295 "independent" or "adult" bears (1957–2005) was 30.5% females and 69.5% males, which does not differ from the sex distribution in the Qaanaaq municipality.

Neither of the two sub-areas (3 and 4) showed a statistically significant difference in the sex ratio in the catch between periods (1957–1990 and 1991–2005; χ^2-tests; P>0.291). There was not any statistically significant difference in the sex distribution between the two areas, in any of the periods ($P > 0.665$); Table 118.

A total of 256 independent polar bears were classified according to age. The percentage of bears that were categorised as "adult" did not differ statistically between the two areas ($\chi^2 = 2.836$; $P = 0.242$, df = 1); Table 116. In contrast to this, the age distribution of the total catch in Upernavik (Areas 3 and 4) differed from the age distribution in Area 3 in the Qaanaaq municipality ($\chi^2 = 8.099$; $P = 0.017$, $df = 1$). This difference is due to the fact that animals described as "old" were more frequent in the catch in Qaanaaqs Area 3 (*i.e.* Qimusseriarsuaq/Melville Bay); Table 116). However, if the categories of "adult" and "old" bears are combined, there is no longer any statistical difference in the respective percentages in Qaanaaq Area 3 and Upernavik Area 3 and 4. Satellite telemetry studies indicate that hunters from these areas all hunt the same population of bears, *i.e.* the "Baffin Bay population" (Taylor *et al.* 2001). In Area 4 there has in recent years been a markedly high occurrence of young male polar bears in the catch (Table 116). In several cases, the hunter in question estimated the age of the bear that was killed. According to these estimates, the "young" bears had an average age of 2.9 years (sd = 0.9; min.-max: 1–5 years; n = 34) and "adult" bears had an estimated average age of 5.0 years (sd = 2.7; min.-max.: 2–11 years; n = 10). There were no estimates as to the individual ages of animals in the age category "old". Thus, hunters in Upernavik tended to estimate the ages of both "adult" and "young" animals a little lower that hunters in the municipality of Qaanaaq.

Size and composition of litters

In the Upernavik municipality, 29 family groups with a total of 52 cubs were reported in the period 1991–2005. On the whole, the percentage of females with cubs was a little lower than it was in the Qaanaaq municipality (Table 115). This was also the case if the whole of the period (1952 to 2005) was taken into consideration (data not shown).

In the cases where hunters specified the age of the cubs (n = 24 litters), 29% of the litters consisted of cubs at around 1 year of age, whereas the remainder comprised litters of 2–3-year-old cubs (in one case, the age was estimated at 2–3 years) (Table 115). Of the 12 cubs for which the sex was specified, 66.7% were males and 33.3% were females. One hunter reported a catch of a female with 0-year-old cubs. This took place in August 1967 to the north of Kullorsuaq.

Discussion

We interviewed polar bear hunters in Northwest Greenland in order to document their observations on changes to the physical environment, the occurrence of polar bears, and the polar bear catch. In this section we discuss the interview survey method and the climate change-related influence on the hunter's environment and the catch.

The survey was originally initiated because both scientific assessments of polar bear populations and reports of a relatively large polar bear catch in Northwest Greenland and along eastern Baffin Island (Canada) implied the populations in these areas were being subjected to over-exploitation (Taylor *et al.* 2003, 2008, Aars *et al.* 2006). There were concurrent reports of changing weather and sea ice conditions, all of which were suspected to impact the occurrence of polar bears.

Hunters in the Qaanaaq and Upernavik municipalities hunt polar bears from two populations: the Ikersuaq/Kane Basin and Baffin Bay populations. Bears that are killed in the Qaanaaq municipality to the north of Innaanganeq/Kap York, and not least in Ikersuaq/Kane Basin itself, belong to the Kane Basin population, whereas those that are taken from the area between Innaanganeq and Sisimiut (ca. 67° N) are part of the Baffin Bay population (*e.g.* Rosing-Asvid & Born 1989, Born & Sonne 2006, this study). Studies involving the use of satellite transmitters have shown that the two polar bear populations are geographically separate (Taylor *et al.* 2001), whereas they are not genetically distinct (Paetkau *et al.* 1999). Based on a mark-recapture study (1991–1998), the Kane Basin population was estimated to be 164 polar bears (95% C.I.: 94–234) while the Baffin Bay population was estimated to be 2074 polar bears (95% C.I.: 1544–2604) (*e.g.* Derocher *et al.* 1998, Taylor *et al.* 2005, 2008, Aars *et al.* 2006). The Baffin Bay population was thought to have decreased to ca. 1600 polar bears in 2004 due to over-exploitation (Aars *et al.* 2006). Since the introduction of the hunting registration system *"Piniarneq"* in 1993, the catch increased in both of the municipalities. In the Qaanaaq municipality the increase was not statistically significant, whereas this was the case in the Upernavik municipality (this study). If estimated population sizes are considered in relation to the reported catch, it is likely that both populations have been, and are still being, subjected to over-exploitation (Derocher *et al.* 1998, Aars *et al.* 2006).

We believe that by interviewing polar bear hunters in Northwest Greenland, we have gathered qualitative data that could not be obtained using other methods. This information supplements the data collected using biological quantitative methods.

Furthermore, the expertise of the hunters and their long-term observations has the potential to provide important details on observed changes in the environment and the polar bear harvest.

The interview survey as a method

The study focused on collecting information factual in nature (sea ice, weather, current patterns, observations of bears, etc.) or data categorised as "Category 1 TEK" (*i.e.* Traditional Ecological Knowledge" about the physical environment; Usher 2000), as opposed to information about cultural values or legal aspects of natural resource exploitation.

Evidently, the interview survey as a method has both positive and negative aspects. One advantage of this method is that the interviews provide a "local" point of view on the subject and therefore facilitate the collection of information on a wide range of issues, such as changes to hunting methods (*i.e.* use of skiffs rather than dogsleds). By interviewing experienced hunters, information can be gathered about changes that have occurred over a number of years, as well as information from observers who have had various degrees of experience and who have different ways of interpreting their observations.

Another obvious advantage of interview surveys is that the act of gathering and conveying the hunters' observations might serve to alleviate some of the frustration that the hunters experience when they feel that they are not being "heard" in the debate on the status of the country's living resources (*cf. e.g.* Tyrrell 2006).

This type of information should be gathered and summarised as systematically as possible in order to minimise the danger of "over generalisation" of the limited quantitative data and/or hypotheses (Wenzel 1999, Usher 2000). A systematic approach to the collection and presentation of data allows the reader to evaluate the background and nature of various statements (*i.e.* their quality, the degree of detail, the validity of field observations as opposed to politically motivated or biased statements, whether the observations are first-hand, a description of what someone else has seen, or "inherited" perceptions, stories or notions).

During the interviews, we attempted to ask a range of detailed questions in a predetermined sequence, in contrast to the "semi-directive" method which makes use of the associative leaps of "ordinary" conversation (*e.g.* Ferguson & Messier 1997, Huntington 1998).

However, the information gathered in interview surveys is influenced by the informants, and to what extent they are a representive selection of polar bear hunters (with regards to both the number of hunters that are interviewed and their individual experience). Thanks to a number of different contacts, we were able to contact almost all of the active and experienced polar bear hunters in the Qaanaaq municipality. In this municipality, the informants represented circa 38% of the registered hunters, and 22 of the 25 were responsible for reporting ca. 71% of the bear catches that were registered in *"Piniarneq"* in the period between 2001 and 2005.

We did not visit Qeqertat and Moriusaq because various pieces of information indicated that the hunters living in these settlements did not engage much in polar bear hunting. Of the total number of polar bears recorded via *"Piniarneq"* during 1993–September 2005 only 2.8% were reported by hunters from Qeqertat and Moriusaq (for the period 2001–2005 this percentage was 1.7).

In the Upernavik municipality, we also predominantly sought out experienced polar bear hunters. Although the 47 informants made up just ca. 15% of the total number of registered hunters in the visited settlements, they had reported ca. 50% of the polar bears that were registered in *"Piniarneq"* during 2001–2005. Several factors prevented us from interviewing polar bear hunters in the settlements Kangersuatsiaq/Prøven and Upernavik Kujalleq/Søndre Upernavik. About 10% of the total number of polar bears reported in *"Piniarneq"* from the Upernavik municipality during 1993–September 2005 (and 2001–2005) were reported from these two settlements. Despite not interviewing hunters from Kangersuatsiaq and Upernavik Kujalleq, we have good reason to assume that the informants from the Upernavik municipality were representative of the experienced polar bear hunters in the area, and that by interviewing them we obtained a general picture of the situation in this part of Greenland.

The estimates of the total annual catch of polar bears in our study are lower than the catch that was registered in *"Piniarneq"*. This is particularly true of the Upernavik municipality. According to our study, the "total" number of bears per year in Qaanaaq in 2004 and the first nine months of 2005 was 72% and 75% of the catch registered in *"Piniarneq"* in those two years, respectively. The corresponding figures in the Upernavik municipality were 28% and 66%, respectively. However, the main purpose of our study was not to gather information on the total number of bears caught nor to evaluate the *"Piniarneq"* reporting system.

All of the interviews started with a set of detailed questions about each of the informant's polar bear catches (Appendix). By asking each hunter specific questions about all of his bear catches, we were able to gain an in-depth insight into the geographic and seasonal distribution of the catch, as well as its age and sex composition. This information, which would not be available from other sources prior to introduction of quotas in 2006, complements the information gathered in response to the more general questions posed during the interviews.

However, the initial questioning regarding each individual bear hunt took a long time and was therefore a tiring process for both the informant and the interviewer. During several of the interviews, the informant was naturally the one who determined the conversation's rhythm, timing, and depth. In a number of cases, this resulted in some of the predetermined questions being left out. Therefore, researchers undertaking interview surveys in the future perhaps ought to consider spending more time on one area and questioning the interviewee on several different occasions. Furthermore, and perhaps more importantly, those undertaking similar surveys ought to consider narrowing down their objectives and asking fewer questions. In our case, the choice could have been between gathering information about either (1) the numerous catches and their details (in order to make a detailed demographic description of the catch) or (2) observations of climate-related changes to the physical environment and the impact that these changes have on polar bears and the catch.

Interview surveys have an obvious limitation as information is to a great extent dependent on the memory of the individual informant. He must possess the ability to observe and combine or reflect upon his observations and a desire or ability to summarise and communicate his knowledge. A consequence of these limitations was that the majority of the data stem from the last 5 to 10 years and are presumably most representative for the most recent period. The same pattern emerged in an earlier interview survey from 1989–90 (Rosing-Asvid & Born 1990). The great and perhaps not unsurprising heterogeneity of the informants' observations and communication skills meant that the degree of detail and apparently the quality of the responses to the different questions varied. It is therefore difficult to judge which of the statements carry the most weight (or to which we ought to attach the most importance). These kind of qualitative data are not easy to summarise. It is particularly problematic to weigh statements when they are contradictory in nature (*e.g.* some informants thought that polar bears have got thinner, others that they are unchanged, and a few that thought that they have got fatter or had changed colour). For a discussion of other aspects of the "pros and cons" of TEK or IQ *cf.* Dowsley & Wenzel (2008). In this summary, we have attempted to find a balance between quantifying the results and a presentation of a selection of the more in-depth individual qualitative statements.

Changes to the physical environment

The characteristics of the sea ice are of fundamental importance to the hunting opportunities of an arctic marine hunting culture. It is evident from the interviews that there has been a change in the sea ice conditions in recent years. The ice tends to form later in the year, has become thinner and on occasion breaks up in the winter. Furthermore, the ice breaks up earlier in the spring. Similar observations have been reported by Canadian Inuit (Riedlinger & Berkes 2001, Nichols *et al.* 2004) – and were also mentioned during the last five or six years by Uusaqqak Qujaukitsoq from Qaanaaq in Huntington *et al.* (2005) (Uusaqqak Qujaukitsoq was not interviewed in our survey). Several of the informants from both municipalities stated that the changes had begun as early as the 1980s – others said that they commenced in the 1990s – but it seemed to emerge that the changes had been most noticeable since the late 1990s. According to a number of informants, the winter of 2005/06 had been more "normal" (until February 2006). The hunter' observations confirmed the satellite based registration which has shown that the sea ice in Baffin Bay and in the North Water area has been in decline since the late 1990s, that the ice has broken up earlier in the spring, and that there has been an overall increase in the number of days with less than 50% ice coverage (Stirling & Parkinson 2006, this study). However, passive microwave satellite data have a resolution of 25 km x 25 km, which means that it is impossible to see details. The hunters' observations are therefore more detailed and evidently provide significant information about what the changes to the sea ice mean for the people and societies that experience them.

In both municipalities informants spoke of a clear recession in the glaciers in recent years. However, some informants indicated that this had been the case as early as the 1980s. A similar decline was reported by Inuit on the eastern coast of Baffin Island (Dowsley 2005, Dowsley & Taylor 2006) and has moreover been registered from satellites since the early 1990s (*e.g.* Comiso & Parkinson 2004, Witze 2008). For the hunters in the Qaanaaq municipality, who regularly use dogsled trails over the glaciers and parts of the Inland Ice Cap, this has been detrimental. The familiar trails are no longer passable and other routes must be found, and the problem is aggrevated by the fact that the reduction in sea ice coupled with its unsafe nature make it highly problematic, and sometimes impossible, for the hunters to travel along the outer coastlines.

In the interviews, it also emerged that the weather had on the whole been more volatile during the 1990s, with rain at unusual times, more unpredictable conditions, and changes to the strength and direction of the wind. Similar changes have been reported by Inuit in the Canadian Arctic (Riedlinger & Berkes 2001, Nichols *et al.* 2004, Huntington *et al.* 2005).

Several hunters remarked that the sea currents had become stronger in recent years. Recordings further south in West Greenland have shown that the sea temperature has consistently been relatively warm since the mid 1990s at all depths, with an increase in temperature from 2000. This implies that there has been an increase in the inflow from the south to West Greenland of water from the relatively warm Irminger Current (Riebergaard & Buch 2005, Riebergaard 2006). It is possible that it is this increased inflow that the hunters in the Qaanaaq and Upernavik municipalities have observed.

Climate change and polar bears

It emerged in the interviews that there have been changes to the sea ice in the past 5 to 15 (20?) years. The extent of the ice in Baffin Bay area has decreased by ca. 8.6% per decade during 1979–2004 (UNEP 2007). The pertinent question is what these changes mean for the polar bears.

Studies in the southwestern part of Hudson Bay (see for example Stirling & Parkinson 2006 and the reference therein) and reflections based on knowledge of the biology and ecological role of the polar bear (Derocher *et al.* 2004, Laidre *et al.* 2008, Wiig *et al.* 2008) imply that it is to be expected that climate change and the ensuing changes to ice and weather conditions will be reflected in changes to the occurrence of polar bears and their physical condition, as well as what they eat. In Hudson Bay, the polar bears now spend more time on land and there has been an increase in the number of "problem bears" observed around the town of Churchill. Female bears have become thinner and there may have been a change in what they eat, shifting from ringed seals to harbour seals (*Phoca vitulina*). Concurrently the polar bear population on the whole has decreased and mortality of young and old polar bears has increased (Stirling & Parkinson 2006 and references therein, Regehr *et al.* 2007).

Studies in Southern Beaufort Sea suggest long-term reductions in sea ice could result in an increasing proportion of the polar bear population on land during the open-water period and an increase in the amount of time individual bears spend on land (Schliebe *et al.* 2008).

A number of the questions in this survey were intended to collect information which may clarify whether such climate-induced changes were also occurring in Northwest Greenland. Most significant in this context was whether the polar bear distribution pattern had changed (if they are closer to the coast due to changes to the ice conditions), whether they have become thinner, and whether they have begun to seek alternative foods.

Changes in distribution of polar bears

The interviews demonstrated that hunters in all areas had observed changes in the occurrence of bears in recent years. Of the 64 hunters who were asked this question, ca. 81% said "yes" and it was apparent that the bears generally occur more frequently in inhabited areas that are regularly used for hunting (11% said "no" and the remainder had no opinion on the matter). In Areas 1–3 (*i.e.* the areas to the north of Nuussuaq) 32% (12/37) of the in-depth responses implied that this was due to an increase in the number of polar bears and this was also implied by ca. 29% (6/21) of the in-depth responses from Area 4 (south of Nuussuaq). On the other hand, in Areas 1–3 circa 24% (9/37) of the affirmative responses indicated that the change in distribution might be due to changes to ice conditions, whereas this explanation was absent from the responses given in the southern part of the Upernavik municipality (Area 4). It is interesting to note that more informants in Upernavik indicated that the change in occurrence is due to an increase in the polar bear population. However, we are unable to offer any immediate explanation for this difference in opinion.

During an interview survey among Inuit in the settlements on the east coast of Baffin Island in 2005, respondents were asked to what extent the polar bear population had increased during the preceding 10–15 years and to what extent bear sightings on land and around inhabited areas had increased in frequency (Dowsley 2005). Most of the informants (15–17 community members were interviewed in each settlement) expressed the opinion that the population had increased in the preceding 10–15 year period. However, some were of the opinion that the observations may reflect a change to the bears' behaviour, and the fact that they occur closer to land as a reaction to the reduction in the sea ice cover, as opposed to an increase in the population (Dowsley 2005, Dowsley & Taylor 2006).

The interviews in Northwest Greenland and Baffin Island showed that the hunters in these areas have noted an increased occurrence of polar bears close to land in recent years. In our study, we deliberately avoided giving any specific year or timeframe in the questions, so that we would not affect the informant's own declaration as to when the changes were observed. However, a number of informants in both municipalities said that the increase in the occurrence began at some point in the 1980s, whereas others indicated that it had taken place during the 1990s.

During the previous interview survey in Qaanaaq municipality in 1989 the hunters did not mention any increase in the occurrence of polar bears in their hunting areas. However, in the interviews from Upernavik in 1990, it emerged that hunters had in part observed an increase in the occurrence of polar bears in the northern part of the Upernavik municipality during the second half of the

1980s, and in part thought that the bears' distribution area (*i.e.* the area in which it was worthwhile hunting for bears) had expanded circa 100 km to the south. As a result polar bear hunting had become a more common occurrence in Nutaarmiut, Tasiusaq and Innaarsuit. Polar bears had also begun to occur on occasion further south (Rosing-Asvid & Born 1990).

The increased catch reported in *"Piniarneq"* since 1993 supports this trend in Upernavik. However, our study indicated that the increased catch to some extent also reflects an increase in the use of skiffs in polar bears hunts. However, it is noteworthy that a similar increase in the polar bear catch has not been observed in the Qaanaaq municipality since 1993 (Fig.1) despite reports of an increased occurrence. The reason for this might be a decrease in the number of active polar bear hunters in this municipality. Our selection of informants in Qaanaaq ensured that by far the majority of the still active bear hunters were interviewed. In 1989 Rosing-Asvid & Born (1990) interviewed a total of 58 hunters, who were thought to comprise almost all of the active bear hunters in the municipality at that time. It is striking that bear hunters under the age of 30 made up 41% of the interviewees in Rosing-Asvid & Born (1990), whereas 100% of those interviewed in 2006 were over 30 years of age. This implies that very few young people in the municipality have become bear hunters in recent years.

It is therefore in our opinion impossible to determine from the interview survey the extent to which an increased occurrence of polar bears in the hunting areas represents an increase in the population or a change in distribution (or for that sake a combination of these factors). However, the interview results (and other sources) do indicate that both the Qaanaaq and Upernavik municipalities have experienced distinct changes in temperature and in sea ice conditions since some point in the 1990s and that these changes may have influenced the distribution patterns of the polar bears. Furthermore, an increase in hunting effort due to an increased use of skiffs for hunting polar bears seems to be a part of the explanation for the increased catch.

Changes in physical condition of polar bears

According to Tyrrell (2006) there is a belief among hunters and elders in Arviat in southwestern Hudson Bay that polar bear numbers are on the increase, and people living there had noted that these bears often looked skinny and unhealthy. However, any particular reason for why bears in this area have become skinny is not given in Tyrrell (2006). During 1980–2004, the total body mass of adult female polar bears from southwestern Hudson Bay decreased, likely reflecting a simultaneous decrease in the duration of the bears' spring feeding season associated with early breakup of the sea ice in Hudson Bay (Stirling & Parkinson 1996).

In our study the hunters were asked whether they had seen any changes to the bears that they had shot. We especially wanted to gather information about the extent to which bears had become thinner. Of 62 hunters who were asked this question, 15 (ca. 24%) thought that the bears' condition had changed (of these 13 thought that they had got thinner, 1 thought that they had got fatter and 1 thought that they had changed colour; the two latter were from Nuussuaq); 43 (ca. 69%) answered "no", or that the polar bears had not changed and the remainder had no opinion on the matter. It is noteworthy that those who expressed the opinion that polar bears have become thinner represented ca. 35% of 37 interviewees in the Qaanaaq municipality/Melville Bay area (*i.e.* Area 1–3). In contrast, all 25 hunters who were asked this question in Area 4 did not think that the bears had changed (1 had no opinion on the matter).

As hunters from both Savissivik and Kullorsuaq catch bears in Qimusseriarsuaq/Melville Bay – with Savissivik hunters tending to operate in the more northerly reaches of the area – it is interesting to note that several informants from Savissivik said that the bears have got thinner, whereas none of the Kullorsuaq hunters expressed the same opinion. This regional variation may either reflect the fact that the polar bears which occur further to the north have become thinner for some undetermined reason – or reflect variation in the individual's subjective assessment of physical condition.

One interpretation of these responses is that polar bears around the North Water (*i.e.* both the Ikersuaq/Kane Basin population and the northern part of the Baffin Bay population) have in recent years tended to be thinner, but that this does not seem to be the case in the southern part of the Upernavik municipality. However, it remains unclear why there should be such a discrepancy.

Even though this implies that some polar bears in the vicinity of the North Water have become thinner in recent years, the interview results on the whole do not indicate that there has been any obvious change to the bears' physical condition.

Changes in diet of polar bears

When asked what the polar bears eat (apart from ringed seals), circa 71% of the respondents mentioned a number of different food items. It is not unusual for polar bears to eat alternative foods (other than ringed and bearded seals) including vegetable foods (grass, plants and seaweed; see for example Winge 1902 and Born & Rosing-Asvid 1989 for a summary of polar bear diet), or to be cannibals (Taylor *et al.* 1985). However, it is conceivable that if polar bears have difficulties finding other food, for example in situations of open water

Bearded seals were also specified as polar bear prey. Photo: Ø. Wiig

or when the duration of the season of open water increases, they might eat more grass and other vegetable foods, become cannibals to a greater extent (Amstrup *et al.* 2006), or may begin to use other food sources such as whale carcasses (Schliebe *et al.* 2008).

Among the informants who provided specific information about alternative foods, ca. 36% (of 70) had observed grass, moss, plants, and seaweed in bears' stomachs. It was striking that 21 of the 25 were from Areas 1–3 and just four were from the southern part of the Upernavik municipality. Whether this is a reflection of an actual regional variation in diet or of differences in the hunters' habits of checking stomach contents is unclear. Some informants mentioned that grass consumption is a phenomenon seen in the spring and summer, and is maybe a result of the bears' hunger. Others said that polar bears have always eaten grass and plants.

In seven cases (four informants from Areas 1–3 and three informants from Area 4) informants stated that polar bears eat other polar bears, on occasion. In six of these cases, the informants specified that it is the larger bears, or males that are cannibals, and that they particularly eat cubs.

Only 4% of the 57 hunters said the polar bears' diet had changed, while ca. 60% were of the opinion that there had been no change (the remainder had no opinion on the subject). The two who answered in the affirmative lived in Savissivik and stated that the polar bears have begun to raid meat depots on land and eat grass to a greater extent. Therefore, the interview results do not on the whole indicate that there has been a change to the polar bears' choice of food in recent years.

Thc catch in figures

Seasonal distribution of the catch

The 1989 interview survey showed that the polar bear catch in the Qaanaaq municipality reached a clear peak in March–April and a less pronounced peak in October–November, which was consistent with the data in "The Hunters' Lists of Game" (HLG; *cf. e.g.* Kapel & Rosing-Asvid 1996 for a description of the HLGs) for the period 1975–1983 (Born & Rosing-Asvid 1989, Rosing-Asvid & Born 1990). The present study confirmed this seasonal pattern in the catch in the period prior to 1991. However, the catch had a much broader seasonal distribution after 1991 (Fig. 33). Interviews with a small number of hunters from the Upernavik municipality in 1990 indicated that the polar bear hunt in that area took place primarily in the period December–April. Data from the HLG during the period 1975–1983 indicated that the polar bear catch in the municipality reached a marked peak in April (Rosing-Asvid & Born 1990). Our study showed a similar peak in the spring for the catch prior to 1991. There was however a broader seasonal distribution after 1991. As in the Qaanaaq municipality, there has also been a tendency towards an increased catch in May and June in the Upernavik municipality in recent years. The interviews indicated that this broader seasonal distribution may represent a generally higher occurrence of polar bears combined with an increase in the intensity of hunting efforts, which is due to an increase in hunting from boats as a result of changed (lighter) ice conditions.

Changes in hunting methods

In our interview survey, the responses clearly indicated that in all the areas investigated there has been a marked increase in the number of polar bears killed from boats since ca. 1990. This increase has been most obvious in the southern part of the Upernavik municipality. A comparison of the geographical location of the individual catches in the periods from 1991–2000 and 2001–2005 respectively revealed that the hunters in the Upernavik municipality still travel west to hunt polar bears. However, whereas hunters used to catch bears from dogsled on the drift ice in Baffin Bay in the early spring (Rosing-Asvid & Born 1990, this study), the hunt in this area has primarily taken place using boats during 2001–2005. The primary reason given for this marked change was that it was difficult or impossible to take a dogsled onto the drift ice because of the unsafe and declining ice conditions. In concert, these same conditions have made it possible for the hunters to more easily search for bears by boat. As some informants pointed out, it is less time consuming to scan large areas for polar bears from a boat than it is on a dogsled.

In an interview survey among polar bear hunters in the Ittoqqortoormiit (Scoresby Sound) in East Greenland, it emerged that there had been a similar marked increase in the percentage of bears killed from boats in the second half of the 1990s, when this period was compared to the 1980s (Sandell *et al.* 2001). These authors ascertained that an increase in the use of skiffs with more powerful motors had allowed hunters to travel around at a faster pace and therefore improved their chances of encountering a polar bear. This resulted in a greater number of bears caught in the season of open water, which has also increased in duration along the coast of East Greenland due to the reduction in sea ice (Sandell *et al.* 2001).

In the Upernavik municipality and to some extent in the Qaanaaq municipality, the effectiveness of polar bear hunting has increased in this way, which is probably a contributing factor to the increased catch.

In the 1980s, some of the spring polar bear hunt in the Upernavik municipality took place out on the drift ice an estimated 120–140 km from the coast (Rosing-Asvid & Born 1990). This pattern also emerged in the present study, in which some of the hunting areas that were pointed out on the maps were as far as ca. 300 km from the coast. We believe, however, that this might be an error. Some hunters undoubtedly travel a long way west when hunting polar bears, but it is hard to believe that a dogsled trip in particular could travel so far to the west. Under favourable conditions it takes a dogsled 5–6 days to travel 300 km out on the pack ice in the west. In these cases, the hunters have probably erroneously placed the kill site too far out to sea on the relatively small, large-scale maps that were used during the interviews.

Geographic distribution

This study indicated that the bear catch in the Qaanaaq municipality since 1991 was geographically distributed over the three sub-areas as follows: (1) ca. 31%, (2) ca. 16% and (3) ca. 53% (Table 111). This distribution is very similar to that which was shown in the interview survey from 1989, when circa 25%, 17%, and 58% of the reported catch (1974–1989) had been made in Areas 1, 2 and 3, respectively (Rosing-Asvid & Born 1990). Qimusseriarsuaq is therefore still the most important bear hunting area in the Qaanaaq municipality. However, it is worth noting that the catch to the north of Ullersuaq/Kap Alexander (Area 1) has in recent years (2001–2005) made up 40% of the total catch. This increase seems to be due to the fact that a relatively large number of bears have been caught from skiffs at the ice edge in the southern part of Ikersuaq/Kane Basin in May–June.

In the Upernavik municipality, the study indicates that in recent years the majority of the polar bear catch has occurred in the southern regions (Area 4) and this represents a combination of an increase in the occurrence of polar bears in coastal areas as well as an increase in hunting efforts.

Age and sex distribution

Our study indicates that female polar bears made up circa 33% of the catch of adult animals for which the sex was specified in the Qaanaaq municipality (2001–2005). There was however regional variation: Adult female bears made up circa 20% of the catch in Area 1 and 21% and 48% in Areas 2 and 3, respectively. These figures and the geographic trends are not significantly different from the data collected from previous interviews and specimen sampling in the Qaanaaq municipality (1982–1996), which showed that adult female polar bears made up 32–33% of the total catch. There was also a predominance of male bears in the catch in Area 1 (22–24% adult females) whereas adult females comprised 32–33% of the catch in Qimusseriarsuaq, *i.e.* Area 3 (Rosing-Asvid & Born 1990, Rosing-Asvid 2002).

Our data from the Upernavik area confirmed the information from the Qaanaaq area by showing that the percentage of adult females in the catch in Qimusseriarsuaq/Melville Bay is high (31% in 1991–2005, 36% in 2001–2005; Table 118) relative to other sub-areas. In Area 4 in the Upernavik municipality, the percentage of adult females was a little lower (ca. 28% in 2001–2005). Interviews with a limited number of hunters in Upernavik in 1990 indicated that in the 1980s, adult female bears comprised 27% of the catch (1976–1990) in this municipality (Rosing-Asvid & Born 1990).

With the introduction of quotas in 2006 it became mandatory for the hunters to report the sex of each bear killed. During 2006–2008 the sex ratio in the catch in Kane Basin was ca. 31.2% adult females and ca. 68.8% adult males. The corresponding figures for the Greenlandic catch of polar bears from the Baffin Bay population during the same period were 33.5% adult females and 66.5% adult males (Born 2009). This confirms the information from the interview survey that there is roughly a 1:2 adult female:adult male ratio in the catch in Northwest Greenland.

Although the geographic distribution of the different age categories in the catch can be subject to a degree of random variation because the number of bear catches was limited (and the information regarding age was based to a great extent on personal judgement) certain trends are worth noting. Information given by hunters in the Qaanaaq municipality suggests that there is a relatively high proportion of old bears in Ikersuaq / Kane Basin and in Qimusseriarsuaq / Melville Bay (the latter was however contradicted by information gathered in the Upernavik municipality). One possible interpretation of the data is that in Ikersuaq and Qimusseriarsuaq, there are older bears, showing some "site fidelity" for good bear habitats (*i.e.* relatively stable ice and a relatively high density of ringed seals; Born & Knutsen 1989, Born *et al.* 1999). Another obvious trend is that in the central area in the Qaanaaq municipality (Area 2) and in the southern part of Upernavik (Area 4), a noticeably large number of young male bears were present in the catch (Table 116). Young males and males in particular have a greater tendency to roam than other age groups (Taylor 1994, Taylor *et al.* 2001). The occurrence of polar bears in these two areas might be more infrequent and it is possible that a higher frequency of stragglers occur in these areas.

Dens and "newborn" cubs

Very few maternity dens were observed in the Qaanaaq and Upernavik municipalities, which is consistent with the information in Rosing-Asvid & Born (1990). Of a total of 41 adult females that were tracked by use of satellite transmitters in Baffin Bay during 1991–1997, none chose to den in Greenland (Taylor & Born unpublished data). However, observations of female bears with very young cubs (*i.e.* that had come out of their den in March–April) in both municipalities do imply that polar bears are denning and giving birth to cubs in the Upernavik and Qaanaaq municipalities, but the extent to which this occurs is probably very limited.

Overall conclusion

The interview survey provided detailed qualitative information on the hunters´ observations of climate-induced changes in their physical environment, occurrence of polar bears, and hunting patterns.

One of the primary results was the observation that a reduction in sea ice quantity and quality has negatively influenced traditional hunting and travelling conditions by means of dogsled. During the same time, the earlier breakup of sea ice and a prolonged open water seasons has led to a change in hunting patterns with an increased use of motorised skiffs in the polar bear hunt.

The survey indicated that there has been an increase in the occurrence of polar bears in the traditionally used hunting areas. However, the information collected during the interview survey was not sufficient to resolve the question of whether an increase in occurrence of polar bears and an increase in the polar bear catch reflect an increase in the exploited population or simply a change in distribution of polar bears caused by a decrease in sea ice.

We conclude that systematic data collection of long-term observations made by experienced hunters may serve as a valuable supplement to scientific studies of arctic wildlife. It would be critical to involve the participation of hunters in both collection and reporting of phenomena related to effects of climate change and occurrence of key hunting and fishing resources.

We recommend "local" observations (or "TEK") be collected systematically through a long-term monitoring system established locally to provide supplementary information on trends in distribution and local density of polar bears.

References

Aars, J., N.J. Lunn & A.E. Derocher 2006. *Polar Bears. Proceedings of the 14th Working Meeting of the IUCN/SCC Polar Bear Specialist Group, 20–24 June 2005*, Seattle, Washington, USA. Occassional Paper of IUCN/SSC No. 32. – IUCN, Gland, Switzerland and Cambridge, UK: 189 pp.

Aastrup, P., C. Egevang, M. Tamstorf & B. Lyberth 2005. Naturbeskyttelse og turisme i Nord- og Østgrønland. – Danmarks Miljøundersøgelser Faglig Rapport No. 545: 133 pp.

ACIA 2005. *Arctic Climate Impact Assessment*. – Cambridge University Press, New York: 1042 pp.

Amstrup, S.C., & C. Gardner 1994. Polar bear maternity denning in the Beaufort Sea. – *Journal of Wildlife Management* 58: 1–10.

Amstrup, S.C., I. Stirling, T.S. Smith, C. Perham & G.W. Thieman 2006. Recent observations of intraspecific predation and cannibalism among polar bears in the southern Beaufort Sea. – *Polar Biology* 29(11), DOI 10.1007/s00300-006-0142-5.

Anon. 1980. Preservation of the Melville Bay, pp. 45–47. In: *Proceedings of the Eighth Working Meeting of the IUCN/SCC Polar Bear Specialist Group, January 1981*. – Gland, Switzerland 1985: 151 pp.

Anon. 1985. Landstingslov nr. 5 af 8. juni 1985 om Grønlands inddeling i landsdele og kommuner. Landstingip inatsisaata tamanut saqqummiunneqarnera/Offentliggørelse af Landstingslov: 9 pp.

Anon. 2006. Pressemeddelelse 18. december 2006 fra Direktoratet for Fiskeri, Fangst og Landbrug (Nuuk).

Anon. 2007. Pressemeddelelse 18. januar 2007 fra Direktoratet for Fiskeri, Fangst og Landbrug (Nuuk).

Anon. 2008. Landstingslov nr. 15 af 5. december 2008 om Grønlands inddeling i landsdele og kommuner.

Barber, D.G., J.M. Hanesiak, W. Chan & J. Piwowar 2001a. Sea-ice and meteorological conditions in northern Baffin Bay and the North Water polynya between 1979 and 1996. – *Atmosphere-Ocean* 39: 343–359.

Barber, D.G., R. Marsden, P. Minnett, G. Ingram & L. Fortier 2001b. Physical processes within the North Water (NOW) polynya. – *Atmosphere-Ocean* 39: 163–166.

Berthelsen, C., I. Holbech Mortensen & E. Mortensen (eds.) 1990. *Kalaallit Nunaat Greenland Atlas*. – Pilersuiffik, Nuuk: 127 pp.

Born, E.W. 1983. *Havpattedyr og havfugle i Scoresby Sund: Fangst og forekomst 1983*. Rapport til Råstofforvaltningen for Grønland og Grønlands Fiskeri- og Miljøundersøgelser fra Danbiu ApS. (Biologiske konsulenter), Hellerup: 112 pp. In Danish with an English summary.

Born, E.W. 1987. Aspects of present-day maritime subsistence hunting in the Thule area, Northwest Greenland, pp. 109–132. – In: Hacquebord, L. & R. Vaughan (eds.). *Between Greenland and America. Cross cultural contacts and the environment in the Baffin Bay area*. Works of the Arctic Centre No. 10. – Arctic Centre. University of Groningen, The Netherlands.

Born, E.W. 1995. Status of the polar bear in Greenland, pp. 81–103. In: Wiig, Ø., E.W. Born & G. Garner (eds.). In: *Polar Bears. Proceedings of the 11th Working Meeting of the IUCN/SSC Polar Bear Specialist Group*. Occassional Paper of IUCN/SSC No. 10. – Gland, Switzerland and Cambridge, UK: 192 pp.

Born, E.W. & A. Rosing-Asvid 1989. Isbjørnen (*Ursus maritimus*) i Grønland: En oversigt. Grønlands Hjemmestyres Miljø- og Naturforvaltning. – *Teknisk Rapport* No. 8: 126 pp. In Danish with an English summary.

Born, E.W. & L.Ø. Knutsen 1989. Observationer af havpattedyr og havfugle i det nordlige Smith Sund, sydlige Kane Bassin og Buchanan Bay, august 1988. – *Grønlands Hjemmestyres Miljø- og Naturforvaltning, Teknisk Rapport* No. 6: 11 pp. In Danish with an English summary.

Born, E.W., Ø. Wiig & J. Thomassen 1997. Seasonal and annual movements of radiocollared polar bears (*Ursus maritimus*) in NE Greenland. – *Journal of Marine Systems* 10: 67–77.

Born, E.W., F.F. Riget, R. Dietz & D. Andriashek 1999. Escape responses of hauled out ringed seals (*Phoca hispida*) to aircraft disturbance. – *Polar Biology* 21: 171–178.

Born, E.W., J. Teilmann, M. Acquarone & F. Riget 2004. Habitat use of ringed seal (*Phoca hispida*) in the North Water area (North Baffin Bay). – *Arctic* 57: 129–142.

Born, E.W. & C. Sonne 2006. Research on polar bears in Greenland, 2001–2005, pp. 135–143. – In: Aars, J., N.J. Lunn & A.E. Derocher (eds.). *Polar Bears. Proceedings of the 14th Working Meeting of the IUCN/SCC Polar Bear Specialist Group, 20–24 June 2005*, Seattle, Washington, USA: 189 pp.

Born, E.W. 2008. *The white bears of Greenland.* – Ilinniusiorfik/Undervisningsmiddelforlag/Education, Nuuk: 128 pp.

Born, E.W., A. Heilmann, L. Kielsen Holm & K. Laidre 2008a. Isbjørne i Nordvestgrønland – En interviewundersøgelse om fangst og klima, 2008. – Pinngortitaleriffik – Grønlands Naturinstitut Teknisk Rapport No. 70: 112 pp. In Danish with an English summary.

Born, E.W., A. Heilmann, L. Kielsen Holm & K. Laidre 2008b. Nannut Kalaallit Nunaata avannaata kitaani – Piniarneq silallu pissusaa pillugit apersuilluni misissuineq, 2006. – Pinngortitaleriffik – Nalunaarusiaq Teknikkimut tunngasoq No. 70: 116 pp. In Greenlandic with an English summary.

Born, E.W. 2009. The catch of polar bears in Greenland, 1993–2008. – Unpublished report to the meeting of the Canadian Polar Bear Technical Committee, 3–6 February 2009, Whitehorse, Yukon, Canada: 6 pp.

Buch, E. 2001. The ocean environment, pp. 111–122. – In: Born, E.W. & J. Böcher (eds.). *The Ecology of Greenland*. – Atuakkiorfik/ Education, Nuuk: 429 pp.

Burns, J.J. 1981. Bearded seal *Erignathus barbatus* Erxleben, 1777, pp. 145–170. – In: Ridgway, S.H. & R.J. Harrison (eds.). *Handbook of marine mammals*. Volume 2 *Seals*. – Academic Press, London, New York, Toronto, Sydney, San Francisco: 359 pp.

Bryder, H., M. P. Porsild & H. Ostermann 1921. Upernivik Distrikt, pp. 430–516. – In: Amdrup, G.C., Louis Bobé, Ad. S. Jensen & H.P. Steensby (eds.). *Grønland i Tohundredeaaret for Hans Egedes Landing*. Volume I. – C.A. Reitzel Boghandel, Copenhagen: 567 pp.

Cappelen, J. 2006. DMI annual climate data collection 1873–2005, Denmark, the Faroe Islands and Greenland – with graphics and a Danish summary. – DMI Technical Report 06–08: 21 pp.

Comiso, J.C. & C.L. Parkinson 2004. Satellite-observed changes in the Arctic.– *Physics Today*. http://.www.physicstoday.org/vol-57/iss-8/938.html

Derocher, A.E., N.J. Lunn & I. Stirling 2004: Polar bears in a warming climate. – *Integrated and Comparative Biology* 44: 163–176.

Dietz, R., C.S. Hansen, E.W. Born, H. Sandell & B. Sandell 2001. Forekomst af "afvigende" isbjørne i Østgrønland. En interviewundersøgelse 1999. – *Faglig Rapport fra Danmarks Miljøundersøgelser*, DMU, nr. 359: 50 pp.

Dowsley, M. 2005. Inuit knowledge regarding climate change and the Baffin Bay polar bear population. – Nunavut Wildlife Research Group Final Report No. 1: 43 pp.

Dowsley, M. & M.K. Taylor 2006. Community consultations with Qikiqtarjuaq, Clyde River and Pond Inlet on management concerns for the Baffin Bay (BB) polar bear population: A summary of Inuit knowledge and community consultations. – Nunavut Wildlife Research Group Final Report No. 2: 83 pp.

Dowsley, M. & G. Wenzel 2008. "The time of the most polar bears": A co-management conflict in Nunavut. – *Arctic* 61: 177–189.

Ehrlich, G. 2006. Grønlands fangere på tynd is. – *National Geographic Danmark* No. 1: 12–27.

Ferguson, M.A.D. & F. Messier 1997. Collection and analysis of traditional ecological knowledge about a population of Arctic tundra caribou. – *Arctic* 50: 17–28.

Ferguson, S.H., M.K. Taylor & F. Messier. 1997. Space use by polar bears in and around Auyuittuq National Park, Northwest Territories, during the ice-free period. – *Canadian Journal of Zoology* 75: 1585–94.

Glahder, C. 1995. *Hunting in Kangerlussuaq. Meddelelser om Grønland, Monographs on Greenland, Man & Society* 19: 1–86.

Glahder, C.M. 2001. Natural resources in the Nanortalik district. An interview study on fishing, hunting and tourism in the area around the Nalunaq gold project. – *National Environmental Research Institute Technical report* No. 384: 1–81.

Glahder, C.M. 2003. Use of local knowledge to mitigate possible conflicts between locals and a new gold mine in south Greenland, pp. 205–214. – In: Udd, J. E. & Bekkers, G. (eds). *Mining in the Arctic. Proceedings of the 7th International symposium.* – Canadian Institute of Mining, Metallurgy and Petroleum, Montréal, Québec, Canada.

Haller, A.A. 1978. *The spatial organization and the marine hunting culture in the Upernavik district, Greenland.* Ph.D. thesis. – The University of Western Ontario, London, Ontario: 382 pp.

Heide-Jørgensen, M.P. & K. Laidre 2006. *Greenland's winter whales.* – Ilinniusiorfik/ Undervisningsmiddelforlag/Education, Nuuk: 122 pp.

Heide-Jørgensen M.P., K.L. Laidre, M.J. Simon, D. Borchers & H. Stern 2008. The effect of sea ice loss on beluga whales (*Delphinapterus leucas*) in West Greenland. – *Polar Research* 29(2), DOI:10.1111/j.1751-8369.2009.00142.x

Holland, D.M., R.H. Thomas, B.D. Young, M.H. Ribergaard & B. Lyberth 2008. Acceleration of Jakobshavn Isbræ triggered by warm subsurface ocean waters. – *Nature Geoscience* 1: 659–664.

Huntington, H.P. 1998. Observations on the utility of the semi-directive interview for documenting traditional ecological knowledge. – *Arctic* 51(3): 237–242.

Huntington, H.P. 2005. Chapter 3 – The changing Arctic: Indigenous perspective, pp. 61–98. – In: ACIA 2005. *Arctic Climate Impact Assessment.* – Cambridge University Press: 1042 pp.

Jakobsen, B.H., J. Böcher, N. Nielsen, R. Guttesen, O. Humlum & E. Jensen (eds.) 2000. *Topografisk Atlas Grønland.* – Det Kongelige Danske Geografiske Selskab og Kort & Matrikelstyrelsen. Copenhagen: 278 pp.

Kapel, F.O. & A. Rosing-Asvid 1996. Seal hunting statistics for Greenland 1993 and 1994, according to a new system of collecting information, compared to the previous Lists-of-Game. – *NAFO Scientific Council Studies* 26: 71–86.

Keith, D., J. Arqviq, L. Kamookak, J. Ameralik & Gjoa Haven Hunters' and Trappers Organization 2005. Inuit Qaujimaningit Nunarnut – Inuit knowledge of polar bears. Gjoa Haven Hunters' and Trappers Organization and CCI Press. – *Solstice Series* No. 4: 242 pp.

Kiliaan, H.P.L., I. Stirling & C.J. Jonkel 1978. Polar bears in the area of Jones Sound and Norwegian Bay. – *Canadian Wildlife Service Progress Report Notes* No. 88, August 1978: 1–22.

Laidre, K.L., I. Stirling, L.F. Lowry, Ø. Wiig, M.P. Heide-Jørgensen & S.H. Ferguson 2008. Quantifying the sensitivity of Arctic marine mammals to climate-induced habitat change. – *Ecological Applications* 18: 97–125.

Nichols, T., F. Berkes, D. Jolly, N.B. Snow & the Community of Sachs Harbour 2004. Climate change and sea ice: Local observations from the Canadian Western Arctic. – *Arctic* 57: 68–79.

Paetkau, D., S.C. Amstrup, E.W. Born, W. Calvert, A.E. Derocher, G.W. Garner, F. Messier, I. Stirling, M. Taylor, Ø. Wiig & C. Strobeck 1999. Genetic structure of the world's polar bear populations. – *Molecular Ecology* 8: 1571–1585.

Perovich, D.K., & J.A. Richter-Menge 2009. Loss of ice in the Arctic. – *Annual Review of Marine Science* 1: 417–441.

Petersen, H.C. 1993a. Upernaviup Kommunia / Upernavik Kommune. Isumalluutinik uumassusilinnik pinngortitamilu eriaqisariaqartunik nalunaarsuineq / Registrering af levende naturværdier i Grønland. – Naqitaq / Rapport No. 14. Published on CD by Namminersornerullutik / Grønlands Hjemmestyre: 47 pp.

Petersen, H.C. 1993b. Avanersuup Kommunia/Avanersuaq Kommune. Isumalluutinik uumassusilinnik pinngortitamilu eriaqisariaqartunik nalunaarsuineq/Registrering af levende naturværdier i Grønland. – Naqitaq/Rapport No. 15. Published on CD by Namminersornerullutik/Grønlands Hjemmestyre: 51 pp.

Ramsay, M.A., & I. Stirling 1990. Fidelity of female polar bears to winter-den sites. – *Journal of Mammalogy* 71: 233–236.

Rasmussen, K. 1919a. Isbjørn – enkens søn, pp. 156–158. – In: *Grønland langs Polhavet. Udforskningen af Grønland fra Melvillebugten til Kap Morris Jesup: Skildring af den II. Thule-ekspedition 1916-18*. Volume I. – Gyldendalske Boghandel. Nordisk Forlag, Copenhagen and Kristiania (Oslo): 288 pp.

Rasmussen, K. 1919b. p. 66. In: *Grønland langs Polhavet. Udforskningen af Grønland fra Melvillebugten til Kap Morris Jesup: Skildring af den II. Thule-ekspedition 1916-18*. Volume I. – Gyldendalske Boghandel. Nordisk Forlag, Copenhagen and Kristiania (Oslo). Bind I: 288 pp.

Rasmussen, K. 1921. Thule Distrikt, pp. 515–567.– In: Amdrup, G.C., Louis Bobé, Ad. S. Jensen & H.P. Steensby (eds.). *Grønland i Tohundredeaaret for Hans Egedes Landing*. Volume I. C.A. Reitzel Boghandel, Copenhagen: 567 pp.

Rasmussen, O.R. 2005. *Analyse af fangererhvervet i Grønland*. – Rapport fra Direktoratet for Fiskeri, Fangst og Landbrug, Nuuk: 158 pp. (Greenlandic version 172 pp.).

Regehr, E.V., N.J. Lunn, S.C. Amstrup & I. Stirling 2007. Effects of earlier sea ice breakup on survival and population size of polar bears in western Hudson Bay. – *Journal of Wildlife Management* 71: 2673–2683.

Riebergaard, M.H. & E. Buch 2005. Oceanographic investigations off West Greenland, 2004. – NAFO Scientific Council Document 05/019: 30 pp.

Riebergaard, M.H. 2006. Oceanographic investigations off West Greenland, 2005. – NAFO Scientific Council Document 06/001: 36 pp.

Riedlinger, D. & F. Berkes 2001. Contributions of traditional knowledge to understanding climate change in the Canadian Arctic. – *Polar Record* 37: 315–328.

Rosing-Asvid, A. & E.W. Born 1990. Fangst af isbjørn (*Ursus maritimus*) i Avanersuaq og Upernavik kommuner: En interviewundersøgelse. – *Grønlands Hjemmestyres Miljø- og Naturforvaltning Teknisk Rapport* No. 23: 63 pp. In Danish with an English summary.

Rosing-Asvid, A. 2002. The polar bear hunt in Greenland. – *Greenland Institute of Natural Resources Technical Report* No. 45: 25 pp.

Sandell, H.T. & B. Sandell. 1991. Archaeology and environment in the Scoresby Sund fjord. Ethno-archaeological investigations of the last Thule culture of Northeast Greenland. – *Meddelelser om Grønland, Monographs on Greenland, Man & Society* 15: 150 pp.

Sandell, H. & B. Sandell 1996. Polar bear hunting and hunters in Ittoqqortoormiit/Scoresbysund, NE Greenland. – *Arctic Anthropology* 33: 77–93.

Sandell, H.T., B. Sandell, E.W. Born, R. Dietz & C. Sonne-Hansen 2001. Isbjørne i Østgrønland: Fangst og forekomst – en interviewundersøgelse. – *Teknisk Rapport* No. 40. Grønlands Naturinstitut. Nuuk: 1–94. In Danish with an English summary.

Schliebe, S., K.D. Rode, J.S. Gleason, J. Wilder, K. Proffitt, T.J. Evans & S. Miller 2008. Effects of sea ice extent and food availability on spatial and temporal distribution of polar bears during fall open-water period in the Southern Beaufort Sea. – *Polar Biology* 31(8), DOI 10.1007/s00300-008-0439-7

Siegstad, H., H.C. Petersen & E.W. Born 2001. Local resource exploitation, pp. 333–346. – In: Born, E.W. & J. Böcher (eds.). *The Ecology of Greenland*. – Atuakkiorfik/Education, Nuuk: 429 pp.

Stirling, I. & C.L. Parkinson 2006. Possible effects of climate warming on selected populations of polar bears (*Ursus maritimus*) in the Canadian Arctic. – *Arctic* 59: 261–275.

Taylor, M. (ed.). 1994. Density-dependent population regulation of black, brown and polar bears. – Ninth International Conference on Bear Research and Management, 23–28 February 1992, Missoula, Montana, U.S.A. *Monograph Series* No. 3: 1–43.

Taylor, M.K., T. Larsen & R.E. Schweinsburg 1985. Observations of intraspecific aggression and cannibalism in polar bears (*Ursus maritimus*). – *Arctic* 38: 303–309.

Taylor, M.K., D.P. DeMaster, F.L. Bunnell & R.E. Schweinsburg 1987. Modeling the sustainable harvest of female polar bears. – *Journal of Wildlife Management* 51: 811–820.

Taylor, M.K., F.L. Bunnell, D.P. DeMaster, R.E. Schweinsburg & J. Smith 1988. ANURSUS: a population analysis system for polar bears (*Ursus maritimus*). – *International Conference on Bear Research and Management* 7: 117–125.

Taylor, M.K., S. Akeeagok, D. Andriashek, W. Barbour, E.W. Born, W. Calvert, D. Cluff, S. Ferguson, J. Laake, A. Rosing-Asvid, I. Stirling & F. Messier 2001. Delineation of Canadian and Greenland Polar Bear (*Ursus maritimus*) populations by cluster analysis of movements. – *Canadian Journal of Zoology* 79: 690–709.

Taylor, M.K., J. Laake, P.D. McLoughlin, E.W. Born, H.D. Cluff, S.H. Ferguson, A. Rosing-Asvid, R. Schweinsburg & F. Messier 2003. Demography and population viability of a hunted population of polar bears. – *Arctic* 58: 203–214.

Taylor, M.K, J. Laake, P. D. McLoughlin, H.D. Cluff, E.W. Born, A. Rosing-Asvid & F. Messier 2008. Demography and conservation of polar bears (*Ursus maritimus*) inhabiting Kane Basin, Canada and Greenland. – *Polar Biology* 31: 491–499.

Teilmann, J., E.W. Born & M. Acquarone 1999. Behaviour of ringed seals (*Phoca hispida*) tagged with satellite-linked radio transmitters in the North Water polynya during fast-ice formation. – *Canadian Journal of Zoology* 77: 1–13.

Tyrrell, M. 2006. More bears, less bears: Inuit and scientific perceptions of polar bear populations on the west coast of Hudson Bay. – Études/Inuit/Studies 30: 191–208.

UNEP 2007. *Global outloook for ice and snow.* – United Nations Environment Programme: 235 pp.

Urquhart, D.R. & R.E. Schweinsburg 1984. *Polar bear. Life history and known distribution of polar bear in the Northwest Territories up to 1981.* – Report from Northwest Territories Renewable Resources. Department of Northwest Territories Renewable Resources, Yellowknife: 69 pp.

Usher, P.J. 2000. Traditional ecological knowledge in environmental assessment and management. – *Arctic* 53: 183–193.

Vibe, C. 1968. Thule-eskimoernes bjørnejagter. – *Tidsskriftet Grønland* No. 6: 175–180.

Wenzel, G.W. 1999. Traditional ecological knowledge and Inuit: Reflections on TEK research and ethics. – *Arctic* 52: 113–124.

Wiig, Ø., J. Aars & E.W. Born 2008. The effects of climate change on polar bears. – *Science Progress* 91: 151–173.

Winge, H. 1902. *Ursus maritimus* L. Isbjørn, pp. 399–409. – In: *Grønlands Pattedyr. – Meddelelser om Grønland, Monographs on Greenland* 21: 320–521.

Witze, A. 2008. Losing Greenland. *Nature* 452: 798–802.

Appendix

Questions – The Polar Bear Catch in the Upernavik and Qaanaaq Municipalities, 2006

General

For each interviewee, the following data is to be recorded: Name, age, address, how long he considers that he has been a hunter, his age, and when he caught his first polar bear.

A summary of the questions

1. *The catch*

The purpose of this part of the interview is to obtain information about: (1) the number of bears shot by each hunter per year (and thereby the total number shot per year in the municipality as a whole, as far back in time as the interviewees can remember), (2) to collect, for each individual bear, data about the location where it was caught, when it was caught, its sex and estimated age (and thereby the catch distribution in terms of age and sex), as well as (3) the circumstances surrounding each individual catch (during a hunting trip by dogsled?, from a boat, etc.)

1.1. (Greenlandic) *Nannunnikuuit? Qassit? Ukioq/ukiut suut?* Have you caught any polar bears (number/year)? Information on each individual bear that the interviewee had shot over the years as far back in time as he could recall. For each bear, it was determined whether the hunter had the main share of the bear or whether he had a lesser part.

1.2. *Nannumik pisaqaqataanikuuit/ammartereqataanikuuit?* Have you been involved in catching/butchering polar bears (number/year)?

For each bear, it was noted:

1.3. Number / sex / age

Sumi pisarineqarpa? Where was the bear caught? (Marked on a map in such a way that we will be able to identify the information about the individual animals at a later date).

Qanga? (Ulloq / qaammat / ukioq). When? (Day / month / year).

Kina ilagalugu pisariviuk? Who were you with?

Nannup suiaassusia? What sex was the bear? (male / female).

Nannup utoqqaassusia? How old was the bear? ("Utoqqaq", "Inersimasoq", "Inuusuttoq"), "Piaraq arnaminik ilalik"; "Old", "adult", "young", "cub with its mother").

Allamik nanoqatiminik ilaqarpa? (Arnaviaq piaralik imaluunniit arnaviaq angutivissamit nuliuffigitittoq). Was the bear with other bears? (With this question, we wanted to obtain information about female bears with cubs or male and female mating pairs).

Qassinik piaraqarpa? How many cubs were there? (In the case of a female with cubs).

Taakku qanoq utoqqaatigippat? (Piaqqisaaruni taakku ukiunik ataatsimiit marlunnut utoqqaassuseqarsinnaapput imaluunniit utoqqaanerusinnaallutik). How old were they? (In the case of a female with cubs).

Piarai ilanngullugit pisarineqarpat? (Aappiuppat kimit?) Were the cubs caught? (If yes, by whom?)

2. Questions 1–31

Hunting and travelling conditions

1. Illoqarfinni / Nunaqarfinni ukioq 2005-imi qassinik nannuttoqar(simasoraajuk?)pa? (ukiuni siuliini eqqaamasinnaasaa malillugu qassit pisarineqarsimanersut apeqqutigineqassaaq)

How many bears do you think were caught in your town/settlement in 2005? (2005 and in each year, as far back as the hunter can remember?).

2. *Kommuuninni nannut qassit pisarineqarsimasorivigit 2005-imi kingumullu?* How many bears do you think were caught in the municipality as a whole? (In 2005 and in preceding years).

3. *Illit nannuttarisartakkatit siornatigornit amerlaneruppat?* Do you catch more bears than you used to?

4. *Maani nunaqarfimmi nanuttarineqartartut siornatigorniit amerlaneruppat?* Are more polar bears caught in this settlement/town than previously?

5. *Nannunniaraangavit/-si aqqutinnaaqarpit/-si?* Do you use regular routes when hunting polar bears?

6. *Nannut nammineerlutik takkuttartut/tikeraartartut amerlanerulerpat?* Do a greater number of polar bears come to visit/come of their own accord?

7. *Aqqutigisartakkat ukiuni kingullerni allanngorsimappat?* Have your routes changed in recent years?

8. *Angallammiit nannuttarisartakkat siornatigumut naleqqiullugit amerleriarsimappat?* Are a greater number of polar bears caught from boats than previously?

9. *Nunami nannuttarisartakkat siornatigumut naleqqiullugu amerlaneruppat?* Are a greater number of polar bears caught on land than previously?

Occurrence of polar bears and dens

10. *Aajangersimasunik nanoqarfeqarpa?* Do the bears occur in particular areas?

11. *Nannut tumaanik sumi takusimavit?* Where have you seen bear tracks?

12. *Nannut aalajangersimasunik aqqutinnaaqarpat?* Do the bears have regular migration routes?

13. *Nanuaqqat tumaanik takusimavit? (ateqqaat tumai).* Have you seen very small bear tracks (*i.e.* from cubs of the year)?

14. Aasakkut nannut sumiittarpat? Where do the bears spend the summer?

15. Nannup illuanik (apissianik) piaqqiviusimasumik takusimavit? Have you seen dens with females with small cubs (maternity dens)?

16. Illumik (apissimik) sinittarfimmik takusimavit? Have you seen temporary dens?

17. Ukiut ingerlanerini illoqarnera (apisseqarnera) allannguuteqarsimava? Have there been changes to the occurrence of dens over the years?

Climate change

18. Sikusarnera allanngorsimava? Have you observed changes to ice conditions?

19. Ilulissat allanngoriartornerannik malugisaqarpit? Have you seen changes to the icebergs?

20. Sermini allanngoriartornernik malugisaqarpit? Have you seen any changes to the glaciers?

21. Apisarnerata allanngoriartorneranik malugisaqarpit? Have you seen any changes to the snow?

22. Sarfap allanngorneranik malugisaqarpit? Have you seen any changes to current patterns?

23. Silap allannguutaanik malugisaqarpit? Have you seen any changes to the weather?

24. Allannguutinik allanik malugisaqarpit? Have you observed any other changes?

Changes to the catch and occurrence of polar bears

25. Allannguutit nannunniartarnermut sunniuteqarpat? Have the environmental changes affected the polar bear hunt?

26. Nannuttarisartakkatit nannullu takusartakkatit allanngorsimappat? Have there been any changes to the polar bears that you have seen or caught?

27. Nanoqassusia allanngorneranik malugisaqarpit? Have you observed changes to the occurrence of polar bears?

The biology and behaviour of polar bears

28. Nuliartunik, nuliuniartunik imaluunniit nuliunermut/nuliarnermut takussutissanik, soorlu tuminik takusimavit? Have you seen mating, mating behaviour or any tracks that indicate these behaviours?

29. Natsertunngikkaangamik nannut sutortarpat? What do the bears eat, apart from ringed seals?

30. Nannunik piniartunik takusimavit? Have you seen polar bears hunting?

31. Nannut nerisartagaat allanngorsimappat? Have there been any changes in what the polar bears eat?

As of 2008 Museum Tusculanum Press has taken over the series
Monographs on Greenland | Meddelelser om Grønland.

Manuscripts should be sent to:

Museum Tusculanum Press
University of Copenhagen
126 Njalsgade, DK-2300 Copenhagen S
Denmark
info@mtp.dk | www.mtp.dk
Tel. +45 353 29109 | Fax +45 353 29113
VAT no.: 8876 8418

Guidelines for authors can be found at www.mtp.dk/MoG

Orders
Books can be purchased online at www.mtp.dk, via order@mtp.dk, through any of MTP's distributors in the US, UK, and France or via online retailers and major booksellers.
Museum Tusculanum Press bank details:
Amagerbanken, DK-2300 Copenhagen S
BIC: AMBK DK KK
IBAN: DK10 5202 0001 5151 08

More information at www.mtp.dk/MoG

About the series
Monographs on Greenland | Meddelelser om Grønland (ISSN 0025 6676) has published scientific results from all fields of research on Greenland since 1878. The series numbers more than 345 volumes comprising more than 1250 titles. In 1979 Monographs on Greenland | Meddelelser om Grønland was developed into a tripartite series consisting of
Bioscience (ISSN 0106-1054)
Man & Society (ISSN 0106-1062)
Geoscience (ISSN 0106-1046)

Hence *Monographs on Greenland | Meddelelser om Grønland* was renumbered in 1979 ending with volume no. 206 and continued with volume no. 1 for each subseries. As of 2008 the original Monographs on Greenland | Meddelelser om Grønland numbering will be continued in addition to the subseries numbering.

Further information about the series, including addresses of the scientific editors of the subseries can be found at

www.mtp.dk/MoG